TECHNICAL REPORT

Bridging the Gap

Developing a Tool to Support Local Civilian and Military Disaster Preparedness

Melinda Moore, Michael A. Wermuth, Laura Werber Castaneda, Anita Chandra, Darcy Noricks, Adam C. Resnick, Carolyn Chu, James J. Burks

Prepared for the Office of the Secretary of Defense

Center for Military Health Policy Research

A JOINT ENDEAVOR OF RAND HEALTH AND THE RAND NATIONAL DEFENSE RESEARCH INSTITUTE

The research reported here was prepared for the Office of the Secretary of Defense (OSD). The research was conducted jointly by the Center for Military Health Policy Research, a RAND Health program, and the Forces and Resources Policy Center, a RAND National Defense Research Institute (NDRI) program. NDRI is a federally funded research and development center sponsored by OSD, the Joint Staff, the Unified Combatant Commands, the Department of the Navy, the Marine Corps, the defense agencies, and the defense Intelligence Community under Contract W74V8H-06-C-0002.

Library of Congress Cataloging-in-Publication Data is available for this publication.

ISBN 978-0-8330-4928-5

Published 2010 by the RAND Corporation
1776 Main Street, P.O. Box 2138, Santa Monica, CA 90407-2138
1200 South Hayes Street, Arlington, VA 22202-5050
4570 Fifth Avenue, Suite 600, Pittsburgh, PA 15213-2665
RAND URL: http://www.rand.org/
To order RAND documents or to obtain additional information, contact
Distribution Services: Telephone: (310) 451-7002;
Fax: (310) 451-6915; Email: order@rand.org

Preface

The Aum Shinrikyo sarin attacks in Tokyo in March 1995, punctuated four weeks later with the bombing of the Alfred P. Murrah Federal Building in Oklahoma City, led U.S. policy-makers to step up systematic disaster preparedness efforts, especially for terrorism. Such efforts accelerated sharply after the terrorist attacks of September 11, 2001, including the creation of the U.S. Department of Homeland Security and a plethora of federal initiatives. Against a backdrop of natural disasters that occur each year in the United States and heightened concern about another influenza pandemic, there is an emerging national consensus that the best path is an all-hazards approach to disaster preparedness planning and that effective local planning is critical.

Federal funding supports preparedness initiatives across cabinet departments, as well as grants to states and certain major metropolitan areas. At the local level, multiple agencies are grappling with a patchwork of federal funding streams and associated grant requirements. Despite clear recognition that disasters occur locally—or at least start that way—most attention to date seems to have been on "top-down" planning from the federal level, representing stovepiped initiatives from different federal agencies. With that in mind, the Office of the Assistant Secretary of Defense for Health Affairs in the U.S. Department of Defense saw an opportunity to strengthen local level disaster preparedness planning by military installations and their civilian counterparts—local governments and local health-care providers, especially the U.S. Department of Veterans Affairs.

This is an interim report for the first phase of a larger study that aims to develop a disaster preparedness support tool for local military and civilian planners. It reflects the formative research carried out from October 2006 through May 2009. It describes the current policy context for domestic emergency preparedness, risk analysis, and capabilities-based planning—the starting points for local planning—as well as results from interviews with local military and civilian planners at five selected sites. All of this information forms the basis for the proposed tool that is described in the final chapter of the report. The next phase of the study will include development and field testing of a proof-of-concept prototype of the tool, which will be supported by funding from the Department of Veterans Affairs.

This report should be of interest to federal, state, and local policymakers and disaster preparedness planners across the range of departments and agencies that have responsibility for domestic disaster management. It should be of particular interest to the Departments of Defense, Veterans Affairs, Homeland Security, and Health and Human Services and the local recipients of their funding and policy guidance. The report should also be of interest to the U.S. Congress and others interested in domestic preparedness that enhances effective and efficient disaster response across the wide range of threats that constitute the new realities of the 21st century.

This research was sponsored by the Office of the Assistant Secretary of Defense for Health Affairs and conducted jointly by RAND Health's Center for Military Health Policy Research and the Forces and Resources Policy Center of the RAND National Defense Research Institute, a federally funded research and development center sponsored by the Office of the Secretary of Defense, the Joint Staff, the Unified Combatant Commands, the Navy, the Marine Corps, the defense agencies, and the defense Intelligence Community.

For more information about the report, contact Melinda Moore or Michael Wermuth. They can be reached by email at Melinda_Moore@rand.org or Michael_Wermuth@rand.org; or by phone at 703-413-1100, x5234 and x5414, respectively. For more information on RAND's Center for Military Health Policy Research, contact the co-directors, Susan Hosek or Terri Tanielian. They can be reached by email at Susan_Hosek@rand.org or Terri_Tanielian@rand.org or by phone at 703-413-1100, x7255 and x5404, respectively. For more information about RAND's Forces and Resources Policy Center, contact the Director, James Hosek. He can be reached by email at James_Hosek@rand.org; by phone at 310-393-0411, extension 7183; or by mail at the RAND Corporation, 1776 Main Street, P.O. Box 2138, Santa Monica, California 90407-2138. More information about RAND is available at www.rand.org.

Contents

APPENDIXES

Figures

Tables

Summary

Local disaster preparedness planners face a major challenge: coordinating and planning with the various civilian entities and military installation counterparts that have authority and responsibility for conducting local response operations. The goal of our overall project is to create a risk-informed planning support tool that will allow local military installations and civilian entities, including local Department of Veterans Affairs health providers—either individually or collectively—to conduct "capabilities-based planning" for local major disasters, with a special focus on the first hours or days after a disaster strikes, when only local response resources will be available. This interim report describes the first phase of our work—development of a framework for a local planning support tool. This work entailed three main steps:

1. Set the current policy framework for local disaster preparedness planning in the United States.
2. Examine what civilian authorities and military installations are doing now with regard to preparedness planning, including their professional connections across local agencies and needs of local planners related to a potential preparedness support tool.
3. From the preceding, derive the design features, components, and data needs for a tool to improve planning at the local level and to design and vet the broad architecture for a capabilities-based planning support tool and complementary tool to enhance local agency connections.

For the first step, we reviewed current policies and programs under which local disaster preparedness now operates and examined the concepts and processes for conducting effective risk assessments and capabilities-based planning, which national policy documents declare should be the standard for preparedness planning. For the second step, we conducted site visits at five locations to learn how communities actually prepare for disasters and to identify the desired features and capabilities for a new planning support tool. We then integrated our understanding of the policy context and of local preparedness planning needs to develop a framework for a local capabilities-based planning support tool.

National Policy Context for Local Disaster Preparedness Planning

Homeland security is often equated with combating terrorism, but the U.S. Department of Homeland Security (DHS) has much broader missions—including federal plans and programs for prevention, protection, preparedness, response, and recovery from all forms of naturally occurring and manmade incidents, both accidental and intentional. Since the terrorist attacks of September 11, 2001, the federal government has been active in providing national guidance and certain standards and practices that can be broadly applied. DHS is one of several federal agencies with authority and responsibility for providing disaster emergency assistance. Others include the Departments of Justice, Health and Human Services, Agriculture, Energy, Veterans Affairs, and Defense.

Two presidential directives provide policy guidance for purposes of this study. First, Homeland Security Presidential Directive 5 (HSPD-5), Management of Domestic Incidents, called for "a comprehensive approach to domestic incident management" (p. 1). HSPD-5 also called for the development of the National Incident Management System (NIMS) (FEMA, 2008b) and the National Response Plan (now known as the National Response Framework, or NRF) (FEMA, 2008a). NIMS provides a standard template for managing incidents at all jurisdictional levels—federal, state, and local—and regardless of cause—terrorist attacks, natural disasters, and other emergencies. The NRF establishes a set of national principles for a comprehensive, all-hazards approach to domestic incident response.

Second, Homeland Security Presidential Directive 8 (HSPD-8), National Preparedness, makes an all-hazards approach to preparedness planning a matter of national policy. The National Preparedness Guidelines (DHS, 2007b) resulting from this directive describe several planning tools; significant among them are the National Planning Scenarios, which cover a broad spectrum of manmade and natural threats; the Universal Task List (DHS, 2007a), which identifies the tasks that need to be performed by all levels of government and from a variety of disciplines to prevent, protect against, respond to, and recover from major disasters and other emergencies; and the Target Capabilities List (DHS, 2007c), which describes core capabilities required to perform critical tasks to reduce loss of life or serious injuries and to mitigate significant property damage.

These policies and doctrines are intended to create a national preparedness system, within which local, state, and federal government entities; the private sector; and individuals can work together to achieve the priorities and capabilities described in these national documents. The local planning support tool that is the goal of our project is based on these policies and doctrines and could play an important role in the national preparedness system.

We examined how guidance contained in the national-level policy documents was (or was not) being operationalized in ways that will support civilian planners at the local level, and how federal programs that are intended to help in local preparedness are structured. Our review highlighted several key points:

- At the state level, the Emergency Management Assistance Compact (EMAC) facilitates resource sharing across state lines during times of disaster and emergency, but localities cannot use the system directly.
- At the local level, there is no single, consistent system or process for mutual assistance.
- Many states have established organizations within the National Guard structure to provide help to localities in an emergency.
- Several federal departments and agencies have preparedness programs that are targeted at the local level. Programs include grants, guidelines, various planning and support efforts, and legal assistance.
- Local planners have access to some tools to support their efforts, but, as of 2008, there was no publicly available decision support tool for local risk-informed capabilities-based disaster preparedness planning—either within government or available commercially.

We also examined statutory authorities and U.S. Department of Defense (DoD) directives and instructions that provide the framework for the role of military installations and other DoD organizations in local preparedness and response activities. Overall, we found that there is ample statutory authority for conducting almost any domestic mission that the military may be called on to perform, especially in the context of domestic disasters or emergencies.

Finally, current national guidance calls for risk assessment and capabilities-based planning within an all-hazards context. Risk analysis is often associated with terrorism planning, but it is not inherently limited to this arena. Risk is composed of three factors: *threat* (the probability of a specified event), *vulnerability* (the probability of damage if an event occurs), and *consequences* (impact on lives and property) (Willis, Morral, et al., 2005). Risk assessment includes assessment of these different components of risk. Risk assessment and capabilities-based planning are both important but not systematically connected in practice. Because DoD asked RAND to address local risk-informed capabilities-based planning, we sought to conceptually connect the two, based on national policy and described methods for each. This would, in turn, guide our development of the planning support tool.

Local Civilian and Military Disaster Preparedness Activities

We conducted site visits at five locations to understand better how communities actually engage in disaster preparedness and to identify the desired features and potential capabilities for a new preparedness support tool. We identified the following patterns across sites.

Civilian Community Networks Were Broader Than Local Military Networks

Across all five sites, civilian and military leaders made fairly consistent distinctions of what the "community" comprised and what constituted their own boundaries for disaster planning purposes. Civilian interviewees generally had a more expansive view of community that included the main city and county and, in some cases, neighboring counties or districts. Their definitions tended to be bounded by where the population lived and worked. Civilians also viewed the installations as largely independent from the city. Military leaders tended to define the community in terms of what they were responsible for in an emergency, which was mostly inside installation boundaries.

Military and Civilians Plan Separately but Use an All-Hazards Approach and Often Participate Together in Exercises

Military installations and civilian planners both tend to approach major disaster planning from an all-hazards perspective. Nevertheless, both the process and the end product may vary by installation and military service and by community. Planning usually starts with a threat or vulnerability assessment. Exercises (tabletops, functional drills, and larger-scale field exercises) are used to test and refine plans and to meet external requirements. Civilian planners rely on several tools to guide the development of plans.

Although there are notable exceptions, military and civilian leaders tend to create their own separate plans in isolation without input from the other party. Once plans are prepared, the level of dissemination and collaboration varies, ranging from simply sharing the plans to participation in joint meetings and exercises.

The level of interaction between military installations and local civilian agencies often depends on the kind of event being planned for and the function of the specific agency. Fire services and public health leaders tend to be more connected across military-civilian boundaries; military and civilian security forces are comparatively less engaged but still cooperate at the tactical level. There is relatively little interaction between civilian planners and the local Department of Veterans Affairs facilities; there is also little interaction with the National Guard in *planning* for disasters. However, the Guard's Civil Support Teams (CSTs) are regularly involved in *responding* to specific emergencies.

Military and Civilian Planners Carry Out Risk Assessment and Capabilities-Based Planning Based on Different Tools Available to Them

Risk assessment for a military installation is a broadly standardized process because all installations are required to meet established DoD benchmarks for antiterrorism protection. However, the process varies substantially across sites depending on the key players. Most of these assessments are not shared with the civilian community. Risk assessments generally take place annually. DoD provides a number of tools to help users conduct risk assessments. The most commonly reported of these is the Joint Staff Integrated Vulnerability Assessment program.

The civilian community has a similar conception of risk assessment, but its process is looser and less standardized. Civilian agencies have also developed a number of tools to address their own needs for a risk assessment template; the Hazard Vulnerability Assessment tool developed by Kaiser Permanente for medical facilities is widely used by civilian planners and by medical personnel on military installations.

We Identified Facilitators and Barriers to Local Disaster Preparedness That Are Important to Consider in Designing a Preparedness Support Tool

Facilitators of local disaster preparedness planning include receipt of external funding, primarily federal government grants passed down through the states; attention from external stakeholders, including federal authorities, the media, and the general public; common guidance, including nationwide, strategic guidance derived from the NRF and from agency-specific guidance from such entities as the Centers for Disease Control and Prevention (CDC); regular interactions among military and civilian stakeholders in the form of meetings and routine or unexpected events; putting faces to names, developing informal connections, and networking; mutual aid agreements, memoranda of understanding, and other formal documentation of the roles and responsibilities of key stakeholders; and information technology.

Barriers to disaster planning include shortcomings in information technologies that preclude essential communications among the many entities involved in a disaster response, especially across military-civilian lines; lack of common terminology; practices for safeguarding information; lack of continuity among the personnel responsible for disaster preparedness or emergency management, especially on the military side; perceived legal constraints that reduce the military's ability to provide support for local disasters; lack of resources; and inaccurate perceptions on the part of both civilians and the military about what each could contribute in the case of a disaster.

Local Emergency Preparedness Networks

We conducted a social network analysis (SNA) to supplement the information we obtained from the interviews at our five study sites. The basic assumption of SNA is that the structure of relationships among a set of actors, and the location of these relationships and actors within a network, have important behavioral, perceptual, and attitudinal consequences for the individual actors and for the system as a whole. We hypothesized that, by seeing their own networks, local emergency response planners might be able to identify missed opportunities for connections.

Most Influential Organizations

Across all five sites, and in both the civilian and military communities, the most influential organizations were consistently emergency management and planning; health and medical; and security, law enforcement, and fire services. These findings align closely with findings from our site interviews.

Communications Flow, Coordination, and Innovation

Our site interviews suggested that fire services and public health organizations are more connected than other organizations across military-civilian boundaries. However, the network analysis suggests that emergency management and law enforcement/security organizations also stand out as particularly influential players that help to improve coordination across each of the larger site-specific networks. These organizations are important not only for increasing coordination *across* civilian and military communities but also for connecting otherwise disconnected organizations *within* either community.

As we learned from the site interviews, there is relatively little interaction between civilian planners and the local Department of Veterans Affairs facilities and little interaction with the National Guard, except for the Guard's CSTs.

Overall, the emergency management networks at all five sites were fairly decentralized and not very densely connected, which means that communications and coordination across the networks are probably less efficient than would be the case in more centrally managed or more densely connected networks. Communications tend to be stovepiped around the larger functional communities in each network, such as the public health/medical community or the law enforcement/security community, but is also concentrated *within* communities, such as the local military installation or the civilian community. Despite this stovepiping, there is a fair amount of regular contact across communities and across functions at various levels; it is just not as common or as frequent as it is within functions and communities.

Resiliency, Redundancy, and Single Points of Failure

A highly centralized network is a less resilient network because the most central node is a potential point of critical failure. The extent of centralization of a network is thus inversely related to its resiliency.

Each of the five study sites is fairly decentralized and, as such, *might* be more resilient to disruption. The site visit interviews suggested that such redundancy exists: several installation emergency managers noted that tactical organizations, such as security forces and fire services, have their own distinct relationships with their civilian counterparts. Our network surveys also reveal some of these relationships.

Organizations that act as brokers between otherwise unconnected pairs of organizations are also potential single points of failure in the network, since they are responsible for bridging the gaps between these otherwise disconnected nodes. Their removal from the network would leave some nodes completely disconnected from one another and, potentially, from the broader network entirely. The organizations playing this broker role and therefore rendering the broader emergency preparedness network most vulnerable were a public health or medical organization (at three of our five sites) and the local emergency management office (at four of the five sites).

Framework for a Local Planning Tool

During our site visits, we garnered information about what civilian and military emergency management personnel would find most useful to support their disaster preparedness efforts. Through a separate review of websites and documents, and complemented by our site interviews, we also inventoried existing preparedness-oriented support tools. We have characterized an inventory of approximately 30 of these, according to such factors as functional support area (risk assessment, planning, event management), access (public versus commercial), hazard(s) addressed, outputs, required user inputs, and target audience.

Existing tools for capabilities-based planning tend to be linked to specific threats or specific localities. The RAND tool will be more broadly applicable—i.e., for all communities and all hazards—and will be automated to alleviate some of the planning burden for local civilian and military planners. The RAND tool would have the following characteristics, based on perceived user needs. It will *leverage existing models and tools* whenever desirable, *automate linkages* for planning activities across disaster phases, and *be applicable to all U.S. communities*, regardless of size. The tool will be *easy to use* and *require minimal technical expertise*. It will be designed in a *Microsoft® Excel® framework*, with which many planners are already familiar. There will be *no barriers to gaining access* to the tool, so it can be widely distributed. The tool will *run on nearly any computing platform*, making it portable.

The tool will *automate use of planning guidance* for civilian and military agencies to help civilian and military officials comply with legal and policy requirements and assist civilian agencies in qualifying for federal grants. We will link recommendations from the tool to their sources for local planners to understand and include a capability for local planners to populate rosters of civilian and military actors within the community who have been identified as being involved with disaster preparedness planning.

The RAND tool will assist local planners by *automating four key outputs*. The tool will (1) automate the process of linking risk assessment to capabilities-based planning, (2) generate resources needed, and, on an optional basis, (3) perform a gap analysis between resources needed and resources available and (4) generate a community disaster preparedness network map, highlighting networking opportunities and a roster of key actors. Figure S.1 shows inputs and outputs for the tool.

Figure S.1
Proposed Inputs and Outputs for the RAND Planning Support Tool

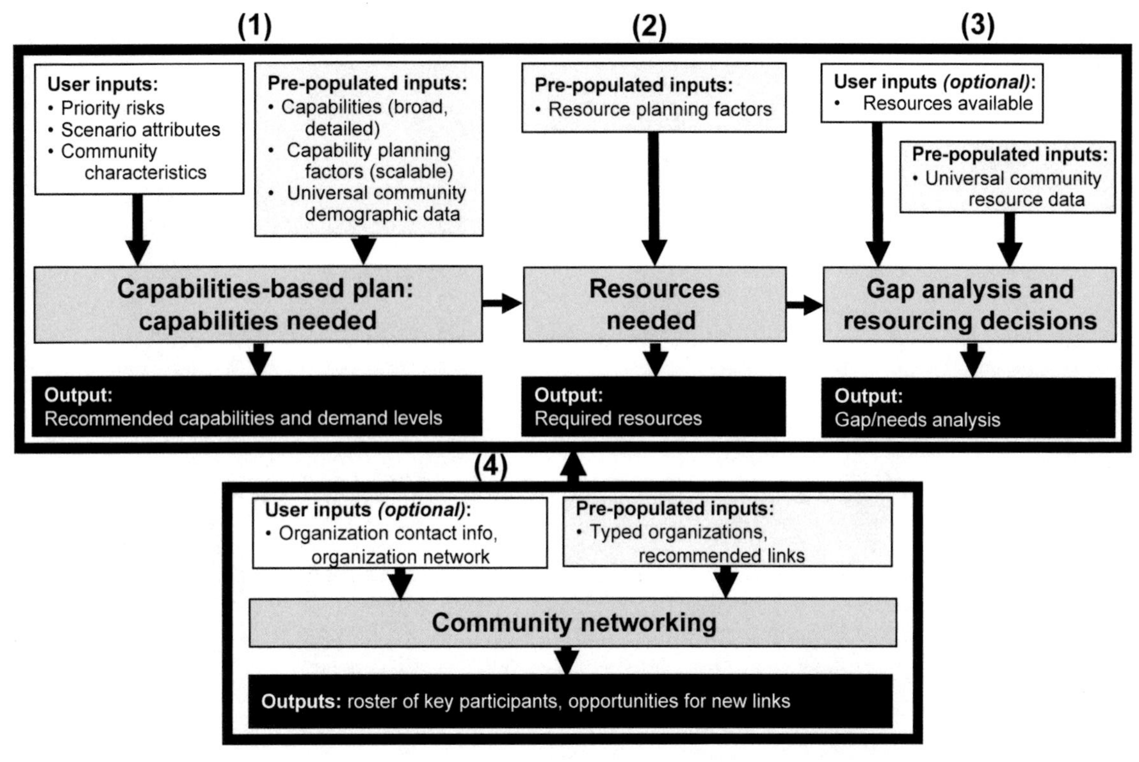

RAND *TR764-S.1*

Based on this framework, our next steps will be to develop and field test a prototype tool that will focus on risk-informed capabilities-based planning. This first prototype will be a "workable" tool—capable of testing inputs and outputs for a subset of capabilities (e.g., medical capabilities) related to some but not all disaster scenarios. We also propose to develop a tool function that can be used to strengthen community networks. The tool will create an environment for community organizations to share contact information; help users identify key organizations with which they can partner to coordinate capabilities should an event occur; and provide an environment in which organizations can share information about upcoming exercises. Although more time and effort will be required to develop a full-scale, fully functional tool that incorporates all capabilities across all scenarios and ready for production and distribution, the next steps in our research effort will be to develop a proof-of-concept prototype tool and field test it in a range of local settings. The proof-of-concept field tests will help in identifying areas for improvement in the further development of the tool and thus will inform the development of the all-capabilities all-hazards planning support tool that is the ultimate goal of this effort.

Acknowledgments

Many people gave generously of their time and expertise in support of this project. Our great appreciation to Deputy Assistant Secretary of Defense for Force Health Protection and Readiness, Ellen P. Embrey, who first had the vision for this project; and Mark Gentilman of the Office of the Assistant Secretary of Defense for Health Affairs, who provided invaluable guidance from the inception of this project to completion of this formative phase.

We thank the numerous civilian and military staff interviewed, including civilian authorities in the San Antonio, Texas, metropolitan area; the cities of Norfolk and Virginia Beach, Virginia; the city of Columbus and Muscogee County, Georgia; the city of Tacoma and Pierce County, Washington; and the city of Las Vegas and Clark County, Nevada; and military staff at Lackland and Randolph Air Force Bases, Fort Sam Houston, Naval Station Norfolk, Naval Medical Center Portsmouth, Naval Air Station Oceana, Naval Amphibious Base Little Creek, Fort Benning, Fort Lewis, McChord Air Force Base, and Nellis Air Force Base. We are grateful for the moral support of officials at the Departments of Veterans Affairs, Homeland Security, and Health and Human Services, who expressed interest in our progress to date and plans for the future.

We thank RAND communications analysts David Adamson and Mary Vaiana for their thoughtful assistance in the structuring of the report, and the report's reviewers—Daniel J. Kaniewski of George Washington University and Ed Chan and Hank Green of RAND—for their comprehensive and thoughtful feedback. We are indebted to our RAND colleagues Susan Hosek and Terri Tanielian, co-directors of the Center for Military Health Policy Research, for their supportive and helpful oversight. Finally, we thank Natalie Ziegler for her help in preparing the final report.

Abbreviations

AAR	after-action report
AETC	Air Education and Training Command
AF/A3	Air Force Deputy Chief of Staff Air and Space Operations
AFB	Air Force base
AFI	Air Force instruction
AHRQ	Agency for Healthcare Research and Quality
AR	Army regulation
ASD (HA)	Assistant Secretary of Defense for Health Affairs
ASD (HD&ASA)	Assistant Secretary of Defense for Homeland Defense and Americas' Security Affairs
ASPR	Assistant Secretary for Preparedness and Response
ATAC	Anti-Terrorism Advisory Council
ATF	Bureau of Alcohol, Tobacco, Firearms and Explosives
ATO	antiterrorism officer
BAMC	Brooke Army Medical Center
BRAC	base realignment and closure
CAMEO	Computer-Aided Management of Emergency Operations
CATS	Consequence Assessment Tool Sets
CAVHCS	Central Alabama Veterans Health Care System
CBIRF	Chemical Biological Incident Response Force
CBMPT	Capability Based Planning Methodology and Tool
CBRN	chemical, biological, radiological, and nuclear
CBRNE	chemical, biological, radiological, nuclear, and high-yield explosives

CCMRF	Chemical, Biological, Radiological, Nuclear, and High-Yield Explosives Consequence Management Response Force
CDC	Centers for Disease Control and Prevention
CEMP	comprehensive emergency management plan
CERFP	Chemical, Biological, Radiological/Nuclear, and Explosive Enhanced Response Force Package
CERT	Community Emergency Response Team
CIA	Central Intelligence Agency
CNI	Commander, Navy Installations
CNIC	Commander, Navy Installations Command
CPS	Continual Preparedness System
CRI	Cities Readiness Initiative
CST	Civil Support Team
DEA	Drug Enforcement Administration
DHS	U.S. Department of Homeland Security
DoD	U.S. Department of Defense
DoDD	U.S. Department of Defense directive
DoDI	U.S. Department of Defense instruction
DOJ	U.S. Department of Justice
DSCA	defense support of civil authorities
DTRA	Defense Threat Reduction Agency
EMAC	Emergency Management Assistance Compact
EMCAPS	Electronic Mass Casualty Assessment and Planning Scenarios
EMP	emergency management program
EMS	emergency medical services
EMWG	emergency management working group
EOC	emergency operations center
ESF	emergency support function
ESSENCE	Electronic Surveillance System for the Early Notification of Community-Based Epidemics
FEMA	Federal Emergency Management Agency

FBI	Federal Bureau of Investigation
FP	force protection
FRP	Federal Response Plan
FY	fiscal year
GIS	geographic information system
HAZUS-MH	Hazards U.S. Multi-Hazard
HHS	U.S. Department of Health and Human Services
HQ	headquarters
HSEEP	Homeland Security Exercise and Evaluation Program
HSPD	Homeland Security Presidential Directive
ICS	Incident Command System
ICT	information and communications technologies
IED	improvised explosive device
IRIS	Incident Resource Inventory System
IT	information technology
JFC	joint force commander
JFHQ-NCR	Joint Force Headquarters National Capital Region
JFO	Joint Field Office
JHOC	Joint Harbor Operations Center
JIATF West	Join Interagency Task Force West
JP	joint publication
JSIVA	Joint Staff Integrated Vulnerability Assessment
JTF-AK	Joint Task Force Alaska
JTF-CS	Joint Task Force Civil Support
JTTF	Joint Terrorism Task Force
LEPC	Local Emergency Planning Committee
LLIS	Lessons Learned Information Sharing
LVMPD	Las Vegas Metropolitan Police Department
MAA	mutual aid agreement
MCRP	medical-contingency readiness plan

MCSO	Muscogee County Sheriff's Office
MEDCOM	U.S. Army Medical Command
MHS	Military Health System
MMRS	Metropolitan Medical Response System
MOA	memorandum of agreement
MOFH	Mike O'Callaghan Federal Hospital
MOU	memorandum of understanding
MRC	Medical Reserve Corps
MSCA	military support to civil authorities
NAB	Navy amphibious base
NAS	naval air station
NCIS	Naval Criminal Investigative Service
NDA	National Defense Area
NDMS	National Disaster Medical System
NGO	nongovernmental organization
NIMS	National Incident Management System
NIMSCAST	National Incident Management System Compliance Assistance Support Tool
NMC	naval medical center
NPG	National Preparedness Guidelines
NRF	National Response Framework
NS	naval station
NWS	National Weather Service
OPNAV	Office of the Chief of Naval Operations
OPTEMPO	operations tempo
OSD	Office of the Secretary of Defense
OSI	Office of Special Investigations
PHEO	public health emergency officer
RMI	Republic of the Marshall Islands
ROC	regional operations center
SAR	search and rescue

SNA	social network analysis
SNS	Strategic National Stockpile
SOP	standard operating procedure
SWAT	Special Weapons and Tactics
TCL	Target Capabilities List
TEEX	Texas Engineering Extension Service
TWG	threat working group
UASI	Urban Areas Security Initiative
UNICEF	United Nations Children's Fund
USAF	U.S. Air Force
USD (P)	Under Secretary of Defense for Policy
USD (P&R)	Under Secretary of Defense for Personnel and Readiness
USNORTHCOM	U.S. Northern Command
USPACOM	U.S. Pacific Command
UTL	Universal Task List
VA	U.S. Department of Veterans Affairs
VISN	Veterans Integrated Service Network
WHO	World Health Organization
WMD	weapons of mass destruction

CHAPTER ONE

Introduction

Local disaster preparedness planners face a major challenge in planning and coordinating across the various agencies that have authority and responsibility for conducting local response operations, including civilian agencies and, in those communities with military installations, their military counterparts. It is one of the great and enduring strengths of the United States that it is organized as a nation with a strong central federal government but comprises 50 sovereign states. However, in the arena of response to and recovery from major disasters and other significant emergencies, this arrangement presents some difficult challenges. Despite the fact that the federal government may often be involved in and has specific capabilities for such response and recovery activities, it is not "in charge" of those efforts for most incidents (except those for which the federal government may have "exclusive" jurisdiction) by virtue of various provisions in the U.S. Constitution—especially the 10th Amendment, Reserved Powers. So local preparedness planners face a major challenge: coordinating and planning with the various civilian entities and military installation counterparts that have authority and responsibility for conducting local response operations.

All disasters and emergencies are local, or at least they start that way. The goal of this project is to create a risk-informed preparedness support tool that will allow local military installations and civilian entities, including Department of Veterans Affairs health providers—either individually or collectively—to conduct local capabilities-based planning for major disasters, with a special focus on the first hours or days after a disaster strikes, when only local response resources are available. The planning will help leaders, both civilian and military, identify gaps in capabilities for purposes of resource allocation and mutual aid.

In this interim report, we describe the first phase of our work—development of a framework for a local planning support tool. This work entailed three main steps:

1. Set the current policy framework for local disaster preparedness planning in the United States.
2. Examine what civilian authorities and military installations are doing now with regard to preparedness planning, including their professional connections across local agencies and needs of local planners related to a potential planning support tool.
3. From the preceding, derive the design features, components, and data needs for a tool to improve preparedness planning at the local level and to design and vet the broad architecture for a capabilities-based planning support tool and complementary tool to enhance local agency connections.

For the first step, we reviewed the current policies and programs under which local disaster preparedness now operates (Chapter Two) and examined the concepts and processes for conducting effective risk assessments and capabilities-based planning (Chapter Three), which national policy documents declare should be the standard for preparedness planning. We wanted to understand the policy context within which local cooperative planning is already taking place and where existing policies facilitate or potentially hinder cooperative planning between the military and civilian sectors. We conducted this review to ensure that the tool we develop is consistent with existing policies and that it does not duplicate effective tools already being used. We also defined the process that we will use to describe risk-informed capabilities-based planning in the development of the planning support tool.

For the second step, we conducted site visits at five locations to learn how communities actually engage in disaster preparedness (Chapter Four) and how they are connected to one another (Chapter Five). These interviews also aimed to identify the desired features and capabilities for a new planning support tool (Chapter Six). Interviews with local civilian officials and installation personnel at these sites, as well as our discussions with relevant officials at the Departments of Defense, Veterans Affairs, Homeland Security, and Health and Human Services, led us to the conclusion that there is no publicly available tool for local risk-informed capabilities-based disaster preparedness planning—either within government or available commercially.

We then integrated our understanding of the policy context and of local preparedness planning practices and needs, to develop a framework for a local capabilities-based planning support tool and complementary tool to assess and improve connections among relevant local agencies (Chapter Six). To complement the report, we include a list of terms and definitions (Appendix A); a more detailed description of disaster preparedness in the civilian sector (Appendix B); relevant policy, doctrine, and organizational entities in DoD (Appendix C); the site visit interview protocols and synthesis guide (Appendix D); detailed summaries from interviews at the five sites (Appendix E); the SNA survey protocol (Appendix F) and detailed findings (Appendix G); and a table of preparedness support tools and methods (Appendix H).

This is an interim report. In the second phase of the study we will develop, field test, and finalize a prototype planning support tool for use by local military installations and civilian authorities, including both local governments and the local service providers from the Department of Veterans Affairs and the Department of Homeland Security Metropolitan Medical Response System. We will prepare a final report, which will be submitted to both our U.S. Department of Defense (DoD) sponsor (which supported the formative research phase reported here) and the sponsor office in the Department of Veterans Affairs (which is supporting the final phase of the project).

CHAPTER TWO

National Policy Context for Local Preparedness Planning

In this chapter, we describe the evolution of authorities, policies, and programs now generally subsumed within the term *homeland security* and the key elements of the policy context for local preparedness planning in the civilian and military sectors. We also refer to three appendixes: definitions and terms (Appendix A), local disaster preparedness in the civilian sector (Appendix B), and relevant DoD policy, doctrine, and organizations (Appendix C).

The Underpinnings of Homeland Security Policy

U.S. homeland security policies and programs provide the overall context for preparedness activities at the local level—both military and civilian. Initially, homeland security activities occurred primarily at the federal level and through federal-to-state programs and resources. However, attention is now focused on a broader set of stakeholders, including not only localities but also private-sector entities—both for profit and not for profit.

The Stafford Act Provides the Statutory Authority for Federal Disaster Assistance to Local Areas, Including Certain Defense Support of Civil Authorities

Historically, most federal response and recovery activities have been in the form of assistance to states and their subordinate localities, predicated on a specific request for such assistance. The best example, and the most widely used authority, for such activities is the Robert T. Stafford Disaster Relief and Emergency Assistance Act (Pub. L. 100-707, 1988, as amended [Stafford Act]; see Appendix A). It is the primary legal basis under which the federal government provides assistance in response and recovery activities to states and localities for major disasters and other emergencies, including terrorist acts.

Since September 2001, much progress has been made in the United States on matters related to homeland security, despite the fact that there is no universally accepted definition of that term. For example, the latest version of the National Strategy for Homeland Security (Homeland Security Council, 2007, p. 3) continues to equate homeland security with combating terrorism:

> Homeland security is a concerted national effort to prevent terrorist attacks within the United States, reduce America's vulnerability to terrorism, and minimize the damage and recover from attacks that do occur.

However, the U.S. Department of Homeland Security (DHS) has important missions much broader than combating terrorism—including federal plans and programs for prevention, protection, preparedness, response, and recovery from all forms of naturally occurring and manmade incidents, both accidental and intentional.

The Homeland Security Council and Several Cabinet Departments Have Critical Roles in Disaster Management

The federal government has been active in providing national guidance and certain standards and practices that can be applied more broadly. Immediately following the September 11 attacks, the President established the Homeland Security Council, a parallel entity to the National Security Council, and the White House Office of Homeland Security (later renamed the Homeland Security Council staff and which is now part of a unified National Security staff) (HSPD-1). In 2002, Congress merged 22 federal agencies and programs to create DHS. But federal assistance to states and localities is complicated by the fact that DHS does not directly control all of the federal homeland security authority or responsibility for providing disaster emergency assistance. Other federal departments with major authority and responsibility include the Departments of Justice (DOJ), Health and Human Services (HHS), Agriculture, Energy, Veterans Affairs (VA), and Defense (DoD).

National Strategies and Presidential Directives Provide Additional Detail to National Guidance for Domestic Disaster Management

A number of other national strategies promulgated during the George W. Bush administration also address various aspects of emergency and disaster preparedness and response activities. They include the National Strategy for Combating Terrorism (NSC, 2006), the National Strategy for Pandemic Influenza (HSC, 2005), the National Strategy for Physical Protection of Critical Infrastructures and Key Assets (DHS, 2003b), the National Strategy for Maritime Security (DHS, 2005), the National Strategy to Secure Cyberspace (DHS, 2003c), the National Strategy for Public Health and Medical Preparedness (Bush, 2007), and the National Strategy for Information Sharing (White House Office, 2007).

In addition to the national strategies, the administration of President George W. Bush issued 24 separate Homeland Security Presidential Directives (HSPDs), spanning a wide range of topics, including terrorist screening procedures, critical infrastructure protection, national preparedness, the protection of food and agriculture, biodefense, maritime and aviation security, and biometrics. As of March 2009, those HSPDs remained in effect.

HSPD-5 states,

> To prevent, prepare for, respond to, and recover from terrorist attacks, major disasters, and other emergencies, the United States Government shall establish a single, comprehensive approach to domestic incident management. The objective of the United States Government is to ensure that all levels of government across the Nation have the capability to work efficiently and effectively together, using a national approach to domestic incident management.

HSPD-5 also called for the National Incident Management System (NIMS) (FEMA, 2008b) and creation of the National Response Plan, which has now been replaced by the National Response Framework (NRF) (FEMA, 2008a), of which NIMS is an integral part.

The NIMS is designed to provide a standard framework for managing incidents at all jurisdictional levels—federal, state, and local—and regardless of cause—terrorist attacks, natural disasters, and other emergencies. By the terms of HSPD-5, states and localities must adopt NIMS to receive federal funding for related emergency and disaster planning, training, and procurement.

The National Response Framework Sets the Context for Coordinated Domestic Response, Including Defense Support of Civil Authorities

The NRF establishes a set of national principles for a comprehensive, all-hazards approach to domestic incident response; however, the principles apply exclusively to response and related preparedness activities, not to prevention, protection, or recovery. The NRF applies those principles, and related roles and structures, across the full spectrum of local, state, federal, and private-sector partners in an effort to foster a coordinated, effective national response. The NRF emphasizes "the importance of planning as the cornerstone of national preparedness" (p. 71).

The National Preparedness Presidential Directive Established the All-Hazards Approach to Disaster Planning Relevant to This Project

Particularly germane to the project described in this report is HSPD-8, National Preparedness. The directive

> establishes policies to strengthen the preparedness of the United States to prevent and respond to threatened or actual domestic terrorist attacks, major disasters, and other emergencies by requiring a national domestic all-hazards preparedness goal, establishing mechanisms for improved delivery of Federal preparedness assistance to State and local governments, and outlining actions to strengthen preparedness capabilities of Federal, State, and local entities.

HSPD-8 makes an all-hazards approach to preparedness planning a matter of national policy. Among other things, HSPD-8 directed the development of a national domestic all-hazards preparedness goal, with

> measurable readiness priorities and targets that appropriately balance the potential threat and magnitude of terrorist attacks, major disasters, and other emergencies with the resources required to prevent, respond to, and recover from them.

The resulting National Preparedness Guidelines (NPG) are intended to do the following:

- Organize and synchronize national (including federal, state, local, tribal, and territorial) efforts to strengthen national preparedness.
- Guide national investments in national preparedness.
- Incorporate lessons learned from past disasters into national preparedness priorities.
- Facilitate a risk-based and capabilities-based investment planning process.
- Establish readiness metrics to measure progress and a system for assessing the nation's overall preparedness capabilities to respond to major events, especially those involving acts of terrorism.

The National Preparedness Guidelines and Associated Documents Establish the Capabilities-Based Preparedness Context Relevant to This Project

The National Preparedness Guidelines define *capabilities-based preparedness* as

> preparing, under uncertainty, to provide capabilities suitable for a wide range of challenges while working within an economic framework that necessitates prioritization and choice. (p. 30)

The National Preparedness Guidelines define *capabilities* as providing

> the means to accomplish a mission or function and achieve desired outcomes by performing critical tasks, under specified conditions, to target levels of performance. (p. 30)

To further the objectives of HSPD-8, the NPG describe several planning tools; significant among them are National Planning Scenarios, the Universal Task List (UTL), and the Target Capabilities List (TCL).

The 15 National Planning Scenarios are designed to highlight the potential scope and complex nature of major disasters and other emergencies. The scenarios cover a broad spectrum of manmade and natural threats, including those involving chemical, biological, radiological, nuclear, explosive, food and agriculture, and cyber terrorism; natural disasters; and pandemic influenza. The National Planning Scenarios are designed to underpin development of standards and metrics for the capabilities required to respond effectively to a wide range of threats. They are specifically designed to be adapted to local conditions.

The UTL identifies the tasks that need to be performed by all levels of government and from a variety of disciplines to prevent, protect against, respond to, and recover from major disasters and other emergencies, both natural and manmade. Developed with extensive involvement from all levels and relevant entities of government and the private sector, the UTL's comprehensive library of tasks ranges from the national or strategic to the incident level.

The TCL describes 37 core capabilities that are required to perform critical tasks to reduce loss of life or serious injuries and to mitigate significant property damage. The TCL supports an all-hazards approach to building specific, identified capabilities that may be needed in the event of terrorist attacks, natural disasters, health emergencies, and other major events:

> The TCL provides a guide for developing a national network of capabilities that will be available when and where they are needed to prevent, protect against, respond to, and recover from major events. These capabilities define all-hazards preparedness and provide the basis for assessing preparedness and improving decisions related to preparedness investments and strategies. (p. 5)

Targeted capabilities may be delivered with any combination of properly planned, organized, equipped, trained, and exercised personnel that achieve the expected outcome.

Together, these policies and doctrines are intended to create a national preparedness system, within which local, state, and federal government entities; the private sector; and individuals can work together to achieve the priorities and capabilities described in these national documents. The local preparedness support tool that is the goal of our project is fully consistent with the guidance and protocols contained in the NRF and in the NPG and other HSPD-8 efforts and could play an important role in the national preparedness system.

Local Disaster Preparedness in the Civilian Sector

We now consider how guidance contained in the national-level policy documents described in the preceding section are (or are not) being operationalized in ways that will support civilian planners at the local level, and how federal programs that are intended to help in local preparedness are structured. We used this information during our site visits to determine whether civilian entities were aware of and participated in existing programs and activities and were involved in collaborative planning with other civilian organizations (see Chapter Four).

Our review of local disaster preparedness in the civilian sector highlighted several key points:

- Several federal departments and agencies, including DHS, VA, HHS, and DOJ, have preparedness programs that are targeted at the local level. Programs include grants, guidelines, various planning and support efforts, and legal assistance.
- Local planners have access to some tools to support their efforts, but there is no publicly available decision support tool for local risk-informed capabilities-based disaster preparedness planning—either within government or available commercially.
- At the state level, the Emergency Management Assistance Compact (EMAC) facilitates resource sharing across state lines during times of disaster and emergency, but localities cannot use the system directly.
- At the local level, there is no single, consistent system or process for mutual assistance.
- Many states have established organizations within the National Guard structure to provide help to localities in an emergency (e.g., the Civil Support Team [CST] and the National Guard Reaction Force).

Federal Structures and Programs for States

Local planners, in both the civilian and military communities, are familiar with many of the national policies and guidelines described in the preceding section, including the NRF, NIMS, and many elements of the NPG, especially the National Planning Scenarios, the UTL, and the TCL. In addition, several federal departments and agencies have preparedness programs that are targeted at the local level—programs that include both federal and nonfederal local entities. These are noted in this section, and each is described in more detail in Appendix B.

Department of Homeland Security. In addition to the national guidance contained in the NRF and the NPG, DHS has a number of assistance and grant programs designed to assist localities:

- Fusion Centers
- Homeland Security Grant Program
- Homeland Security Exercise and Evaluation Program
- Urban Areas Security Initiative (UASI)
- Citizen Corps
- Metropolitan Medical Response System (MMRS).

For more detail on each of these programs, see Appendix B.

Department of Veterans Affairs. The VA has issued the comprehensive *Emergency Management Program Guidebook* addressing all hazards (EMP). The guidelines require each VA facility—both health units and other offices—to do the following:

- Complete a nine-step process that includes a vulnerability assessment across a range of hazards and development of an emergency operations plan.
- Coordinate its efforts in an ongoing manner with community response partners (other health-care organizations, local and state government, suppliers and nongovernmental organizations, such as the Red Cross); however, the guidebook provides little specific guidance about how coordination should be achieved, and, except for minor references involving notifications and exercises, coordination with community partners does not appear in the performance-evaluation section.

Department of Health and Human Services. HHS has primarily focused on providing public health and medical support, including countermeasures, in disaster events and developing guidance for public health and medical responses, including medical treatment in mass casualty events. Four entities in HHS have major responsibilities, programs, and resources in this regard:

- Assistant Secretary for Preparedness and Response (Hospital Preparedness Program; National Disaster Medical System, including Disaster Medical Assistance Teams, the Disaster Mortuary Operational Response Team, and National Nurse Response Teams)
- Office of the Surgeon General (Medical Reserve Corps)
- Centers for Disease Control and Prevention (CDC) (Cooperative Agreement on Public Health Emergency Preparedness, Strategic National Stockpile [SNS], Cities Readiness Initiative [CRI])
- Agency for Healthcare Research and Quality (AHRQ) (public health emergency management resources, including tools and resources related to community planning, mass prophylaxis, modeling, pandemic influenza, pediatrics, and surge capacity; evidence reports; and notes from selected meetings and conferences).

Department of Justice. DOJ has numerous activities directed at preventing terrorism and responding to attacks that may occur. Two that are relevant to this project are

- Federal Bureau of Investigation (FBI) Joint Terrorism Task Forces
- U.S. Attorneys' Anti-Terrorism Advisory Councils.

State Structures

Many states have programs designed to provide assistance directly to localities for major disasters and other emergencies. Those programs differ significantly from state to state. However, several state-level structures are essentially standard nationwide, including mutual-assistance agreements and National Guard organizations (see Appendix B for more details on each of these).

Mutual Assistance. All 50 states, as well as the District of Columbia, Puerto Rico, and the U.S. Virgin Islands, have enacted legislation to become part of EMAC, a mutual aid compact that facilitates resource sharing across state lines during times of disaster and emergency.

EMAC has been used extensively in major disasters. There is no standard, national system or process for mutual assistance at the local level.

National Guard Organizations. Several states have established organizations within the National Guard structure that have been supported with direct appropriations from Congress to provide assistance at the state and local levels for various types of incidents. Those relevant to our study are as follows:

- CST
- Chemical, Biological, Radiological, Nuclear and High-Yield Explosive (CBRNE) Enhanced Response Force Package.

These are described in more detail in Appendix B.

Department of Defense Authorities for Civil Support and Preparedness Activities

Specific national policy documents, statutory authorities, and DoD directives and instructions provide the framework for the role of military installations and other DoD organizations in local preparedness and response activities. In our site visits to military installations, we assessed whether interview participants were aware of these DoD policies. We also gathered information to understand whether and how installations were implementing specific requirements and existing programs and activities, especially in collaborative planning with civilian organizations (see Chapter Four).

In this section, we briefly review the basic constitutional and statutory foundation for the military's domestic mission. Overall, our review indicates that there is ample statutory authority for conducting almost any domestic mission that the military may be called on to perform, especially in the context of domestic disasters or emergencies. Most military operations in support of civil authorities will likely be for Stafford Act–type incidents (i.e., major domestic disasters and emergencies), but DHS and other federal agencies (and, through them, states) can request other types of military assistance, supported by a number of statutory authorities. Although the Posse Comitatus Act (18 U.S.C. §1385) and other provisions do place some restrictions on some specific law enforcement activities, in certain circumstances, the military may use traditional battlefield or more strategic intelligence, surveillance, and reconnaissance capabilities, including several specific law enforcement authorities (described in this section), to assist civil authorities.

National Strategy for Homeland Security

The National Strategy for Homeland Security emphasizes DoD's important role in civil support:

> While defending the Homeland is appropriately a top priority for the Department of Defense, the country's active, reserve, and National Guard forces also must continue to enhance their ability to provide support to civil authorities, not only to help prevent terrorism but also to respond to and recover from man-made and natural disasters that do occur. (HSC, 2007, p. 51)

In recognition of its responsibilities, DoD has promulgated its own Strategy for Homeland Defense and Civil Support (herein, the HD/CS), which provides significant policy direction for the civil support mission.

The foundation for the military's role in supporting civil authorities and responding to disasters has been established through a long history of law and policy. This history both empowers the military to respond and restricts certain activities. In our review, we focus on a few key topics: the constitutional basis for civil support, the Posse Comitatus Act, and the statutory authority for domestic missions.

Constitutional Basis for Civil Support

DoD authority for civil support is grounded in the U.S. Constitution. Article One gives Congress the power to create military forces and provide for their regulation, and contains explicit language for "calling forth the militia" to enforce laws and suppress rebellions and insurrections. Article Two designates the President as Commander in Chief not only of regular federal forces but also of the state militias, when in federal service—*militia* being what we now know as the state National Guard. Article Four states that the United States shall protect each of the states not only against invasion but also against "domestic violence."

Posse Comitatus Act

In the first century of the republic, there were a number of instances in which the military was used to enforce laws, which gave rise to some criticism of those activities—most particularly, military actions in the reconstruction and post-reconstruction periods in the South. It was the latter circumstances that caused Congress, in June 1878, to pass what has come to be called the Posse Comitatus Act (18 U.S.C. §§1385, 1878). *Posse comitatus* translated from Latin means "the power or force of the county."

The Congress did not proscribe the use of the military as a "posse comitatus" or otherwise as a means of enforcing the laws in U.S. Code Title 10, Armed Forces (which describes the organizations and authorities of the various DoD entities). Rather, it made it a crime under Title 18, Crimes and Criminal Procedure, to willfully use "any part of the Army or the Air Force as a posse comitatus or otherwise to execute the laws." In the same act, the Congress created a very broad exception to the application of the act for those "cases and under circumstances expressly authorized by the Constitution or Act of Congress."

Since the enactment of the Posse Comitatus Act, the Congress has created a number of statutory exceptions to that act, which fall into four major categories: disaster relief, insurrections and civil disturbances, counterdrug operations, and counterterrorism and weapons of mass destruction.

Statutory Authority for Domestic Missions

There is extensive statutory and regulatory authority for DoD to provide what is now collectively referred to as defense support of civil authorities (DSCA). The statutes and authorities include the Stafford Act, the provisions of law collectively referred to as the Insurrection Act (10 U.S.C. §331–335), as well as some specific statutory authority for dealing with nuclear, chemical, and biological terrorism.

Robert T. Stafford Disaster Relief and Emergency Assistance Act. The best example and the most widely used authority for civil support activities is the Robert T. Stafford Disaster

Relief and Emergency Assistance Act (Stafford Act), which provides for federal assistance to states and localities for certain major disasters and emergencies.

Activating aid under Stafford usually requires a request from a governor to the president; requests for federal assistance in major disasters require a declaration by the president. If the president approves the request, he or she will direct the Secretary of Homeland Security to provide federal assistance, and then a further request may be made to the Secretary of Defense for support from DoD assets. Most authorities require that DoD be reimbursed.

However, the federal government, including DoD, may conduct activities—even in the absence of a request—when immediately necessary to protect people from death or serious injury. Local commanders are authorized to undertake "immediate response," notwithstanding the absence of direction from higher authorities, to respond to requests from civilian entities to "save lives, prevent human suffering, and mitigate great property damage" (DoDD 3025.1; JP 3-28).

There is no standard process for the "reverse" of DSCA—civilian support to military entities.

Insurrection Act. Provisions of Title 10 have come to be referred to as the Insurrection Act (10 U.S.C. Chap. 15)—a direct reference to the provisions of Article One of the Constitution. Those provisions authorize the use of military forces to suppress insurrections against state governments; to suppress "unlawful obstructions, combinations, or assemblages, or rebellion against the authority of the United States"; and to suppress, in a state, any

> insurrection, domestic violence, unlawful combination, or conspiracy, if it (1) so hinders the execution of the laws of that state, and of the United States within the state, that any part or class of its people is deprived of a right, privilege, immunity, or protection named in the Constitution and secured by law, and the constituted authorities of that state are unable, fail, or refuse to protect that right, privilege, or immunity, or to give that protection; or (2) opposes or obstructs the execution of the laws of the United States or impedes the course of justice under those laws. (10 U.S.C. §333)

The Insurrection Act has been invoked on numerous occasions, most notably for the integration of schools in the South and for the riots in major U.S. cities in the 1960s. It was used more recently in connection with the riots sparked by the Rodney King incident in California in 1992 (10 U.S.C. §§331–335).

Statutory Authority for Dealing with Nuclear, Chemical, and Biological Terrorists

Title 10, Section 382 of the U.S. Code authorizes the Secretary of Defense, at the request of the U.S. Attorney General, to provide assistance for certain emergencies involving a biological or chemical "weapon of mass destruction," including activities to "monitor, contain, disable, or dispose" of the weapon. Although there is a general prohibition in this section against the military engaging in arrest, search, seizure, and the "direct participation in the collection of intelligence for law enforcement purposes," a later clause specifically authorizes those types of activities by the military if a determination is made that such action is required for "the immediate protection of human life, and civilian law enforcement officials are not capable of taking the action," or authorized by other laws.

In a parallel statute involving nuclear materials, the Secretary may provide assistance at the request of the Attorney General to enforce that criminal statute, and—with an explicit exemption to the Posse Comitatus Act—such assistance may, under certain circumstances, include arrests, searches, seizures, and other activities "incidental to the enforcement of this section, or to the protection of persons or property from conduct that violates this section" (18 U.S.C. §831).

In addition to the Stafford Act, many of the statutory authorities require, or, at a minimum, authorize, the Secretary of Defense to seek reimbursement from the entity that requests the assistance (e.g., 10 U.S.C. §382), and many also direct that the Secretary determine that the requested support will not adversely affect readiness or other military operations (e.g., 18 U.S.C. §831).

These authorities for the military's domestic mission are reflected in DoD policy and doctrine. Appendix C to this report contains more detailed information on other aspects of DoD policy and doctrine for civil support and homeland defense, including a summary of the 2006 Quadrennial Defense Review (DoD, 2006b); applicable Office of the Secretary of Defense (OSD) directives and instructions; joint publications; applicable military department regulations, directives, and instructions; and information on DoD entities—military and civilian—with authority and responsibility in this area.

Summary

The Stafford Act and other statutory authorities have long provided the basis for federal assistance to states and localities for most forms of disasters—natural and manmade. Since the September 11 attacks, new national strategies and presidential directives and other major policy documents have established a robust system, not only at the federal level but also with broader application nationwide, for prevention, protection, response, and recovery for numerous scenarios. As noted, that system includes critical activities for preparedness, including risk analysis for capabilities-based planning, which is the subject of the next chapter.

CHAPTER THREE

Risk-Informed Capabilities-Based Planning

Guidance contained in national policy documents that address preparedness activities (e.g., the NPG and NRF) suggest that the standard should be risk-informed capabilities-based planning at all levels of government. This chapter describes the concepts and processes for conducting effective risk assessments, and defines and describes capabilities-based planning. We then describe the process that we will use for risk-informed capabilities-based planning in the development of the planning support tool.

What Is Risk Analysis?

Individuals and institutions expose themselves to the potential for harm with almost any activity they undertake. This simple premise serves as the basis for the concept of risk. In its narrowest sense, risk is a function of probability and consequences (see Figure 3.1). Through this lens, those events that are more likely and have greater consequences represent the greatest risks.

The earliest modern applications of risk analysis adopted this narrow approach, considering risk to be only what could be assessed using methods of classical probability theory (Knight, 1921). That is, risk analysis was viewed as appropriate only for assessing events that could be observed and analyzed in terms of frequency. However, many modern applications of risk analysis do not fit this mold. The data used in risk analysis to set event probabilities that inform, for example, the design of levees and nuclear reactors, and schedules for deliveries of goods to stores, are often sparse, or the underlying processes are dynamic, thus limiting one's ability to draw useful inference from historical data. As a result, risk analysis has developed to recognize probability as more than what can simply be empirically observed. Instead, probability is treated as representing degrees of belief that events will occur. Thus, intuition and judgment are fundamental components of risk (Keynes, 1921).

In addition, more modern applications of risk analysis view consequences as more than what can be tangibly observed. The most conventional types of consequences considered in risk analysis include loss of property (both infrastructure and wealth) and loss of life and vitality (i.e., fatalities and injuries). However, studies of how risks are perceived have demonstrated that judgments of risk are based on more than statistical counts of these consequences. People care about the context within which events occur. For example, people tend to view hazards that are involuntary, catastrophic, delayed, and poorly understood as presenting greater risks than those that are voluntary, chronic, immediate, and well understood (Slovic, Fischhoff, and Lichtenstein, 1979). In addition, how people view and process hazards in these terms is defined not only by science but also by the social amplification of risk that can occur as information

Figure 3.1
Elements of Risk

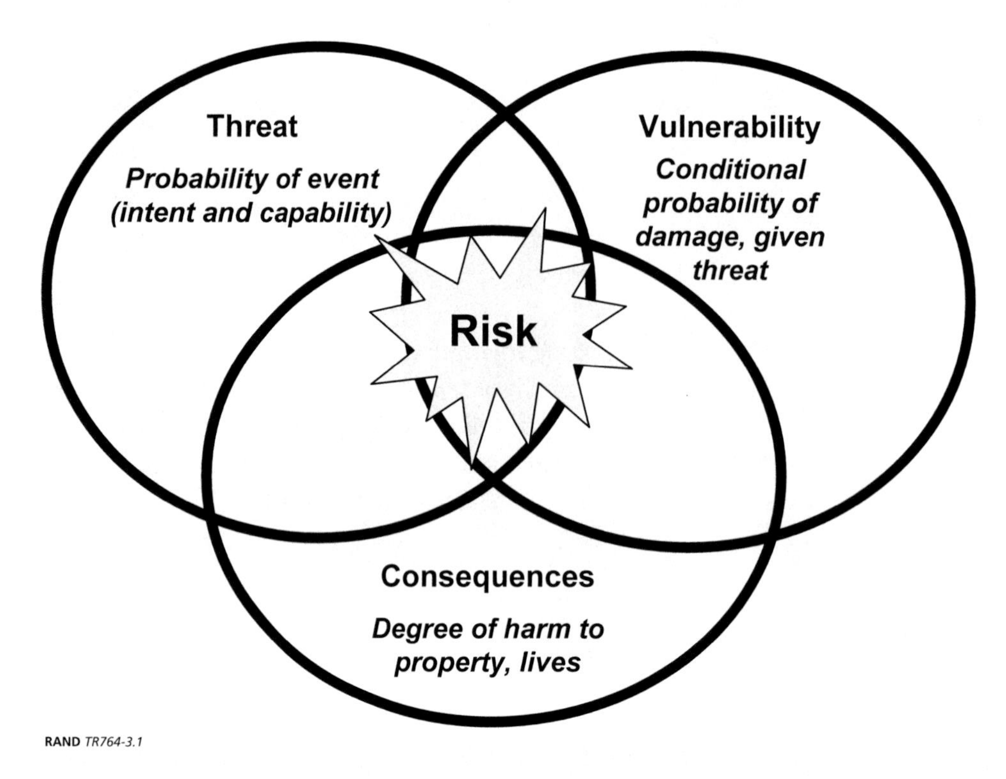

RAND *TR764-3.1*

about hazards is processed by individuals, organizations, the media, and society (Kasperson and Kasperson, 1996).

In the end, all of these factors play into how people judge risks, acceptability of risks, and worthiness of risk management alternatives (Fischhoff et al., 1981). Thus, risk is most accurately viewed not as an actuarial exercise, but instead as a social construct: a combination of risk assessment data and stakeholders' views regarding what they perceive to be most important risks and most important ways to address them. It is entirely possible that a local community may be particularly fearful of a rare but serious event and consider it to be among their priorities, even if the "numbers" do not point to such an event as a high risk. The risk assessment process, in turn, feeds into local planning, to address the most important local risks.

Applying Risk Analysis: The Terrorism Example

As described earlier, intentional, manmade disasters—especially terrorism—comprise one important group of disasters that are subject to federal, state, and local preparedness planning. We will use the terrorism example to discuss risk assessment in more detail.

Adopting the framing of risk analysis as described in the preceding section presents several challenges for assessing risks presented by terrorism. These can be recognized by considering the three elements that contribute to terrorism risk: *threat*, *vulnerability*, and *consequences* (see Figure 3.1). Each of these factors has been defined from the perspective of risk analysis (Willis, Morral, et al., 2005).

Threat exists only when a person or group has both the intention to cause harm and the capabilities and wherewithal to accomplish their intended actions. Thus, threat is a manifestation of terrorists' motivations, objectives, resources, and skills. Threats are relevant only if directed toward targets that are vulnerable to the specific mode of attack at the time it is attempted.

Vulnerability is a property of the system being considered; it is determined by how infrastructure is designed and operated, the environment within which it exists (shaped by, for example, terrain or security postures), the time that an event occurs (for example, day versus night or summer versus winter), and the conditions that exist at that time (for example, inclement weather or heavy traffic).

Consequences are inherently multidimensional and may be far reaching and distributed over time, places, and sectors of the economy. This makes their specification particularly daunting. The direct consequences of terrorism come to mind quickly. On September 11, 2001, more than 3,000 people died, many others were injured, and tens of billions of dollars of property was destroyed. However, the consequences of terrorism extend beyond these direct effects. In addition to the common impacts on property and human life, terrorism can affect stability in governance, public confidence, and national security.

Consequences can also propagate through interconnections of infrastructure and social amplification of risk through behavioral systems. For example, consider how disruptions to power supply can trigger disruptions to all industries that rely on power or how the reactions to terrorism led some people to change vacation plans or choose not to fly in the months following September 2001. Although the existence of such indirect effects is recognized, they are not well understood, and little has been done to begin the discourse necessary to establish the importance of each category of consequences relative to the others. Indirect effects are another good argument for considering perceived, not just objective, risk.

Finally, terrorism risk exists only when all three of these factors—threat, vulnerability, and consequences—are present. For risk to exist, a person or group must have the capabilities and intent to present a threat of attack on a vulnerable target in a manner that would have consequences of concern to those who may be at risk (e.g., the citizens of the United States).

Connecting Risk Assessment to Risk Management

Risk analysis is generally viewed as encompassing two distinct processes: risk assessment and risk management (Renn, 2005). *Risk assessment* refers to efforts required to understand the nature and extent of risks, as well as how those risks are perceived. *Risk management* refers to the complementary efforts to decide how to respond to those risks.

Despite being distinct in terms of their objectives, risk assessment and risk management are inherently interdependent. Pre-assessment of risks is an important factor in determining what risks will be managed, as well as which will be appraised. Similarly, detailed characterization and evaluation of risks can lead to changes in priorities for both appraisal and management of risks.

In the context of terrorism, a challenging aspect of risk management stems from the possibility of modifying risks at several points within the causal chain of events that leads from a hazardous activity to a harm. Figure 3.2 illustrates a taxonomy described for classifying the

Figure 3.2
Taxonomy of Four Alternative Approaches to Terrorism Risk Management

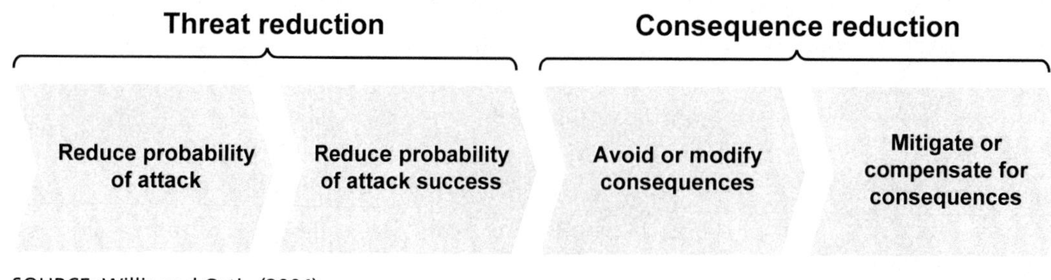

SOURCE: Willis and Ortiz (2004).
RAND *TR764-3.2*

different approaches for risk management for terrorism (Willis and Ortiz, 2004). Following this taxonomy, approaches for managing risks fall into two broad categories: threat and consequence reduction.

Threat Reduction

Threat reduction efforts focus on reducing the probability that successful attacks occur; these approaches aim to reduce the probability of attack and the probability that an attack will be successful. Threat reduction can be accomplished in three ways. The first two are aimed at reducing the probability of an attack, and the third is aimed at reducing the likelihood of success if an attack occurs:

- First, security efforts may deter terrorists from attempting attacks. For example, if terrorists perceive border inspections as posing too great a risk of detection, they may decide not to attack.
- Second, security measures may change terrorists' capabilities to attack. For example, it has been suggested that U.S. disruption of al Qaeda training camps in Afghanistan made it more difficult for terrorists to plan, train for, and launch sophisticated attacks.
- Third, security measures can make it more likely that attempted attacks would be thwarted before they could be completed. The successful interdiction of plots to attack Fort Dix and JFK International Airport are examples of how targeted intelligence efforts can reduce threats in this way.

Consequence Reduction

Risks can also be reduced by changing the impact of consequences on targeted areas. One approach (third block in Figure 3.2) addresses potential consequences before an event, and the other (fourth block in the figure) seeks to reduce or compensate for consequences after an event. The first approach is to implement measures that change where attacks occur or how significant the consequences are. For example, a study of Los Angeles International Airport suggested that the risks of small bomb attacks could be reduced by checking people in for flights more quickly, thus reducing waiting lines and the density of people in areas that might be at risk of bombings (Stevens et al., 2004). The second approach is to manage risks by reducing the burden of impacts on those affected after events occur. Examples of policies based on this approach include the September 11th Victim Compensation Fund (DOJ, undated) and the

Terrorism Risk Insurance Act of 2002 (Pub. L. 107-297) implemented following the attacks of September 11, 2001 (Dixon and Stern, 2004).

DHS and DoD efforts to manage terrorism risk have recognized these four approaches and generally have adopted a layered approach in case any single approach fails. As a result, evaluation of any single risk management effort must be assessed within the context of the entire risk management portfolio of which it is part. This type of analysis is analytically challenging. Capabilities-based planning provides one approach for addressing this challenge (see Figure 3.3).

Capabilities-Based Planning

As described earlier in this report, national-level guidance promulgates capabilities-based planning. In this section, we describe such planning in the military and civilian contexts.

In the military context, capabilities-based planning is intended to replace threat-based planning of the type that was prevalent during the Cold War.

Capabilities-based planning is "planning, under uncertainty, to provide capabilities suitable for a wide range of modern-day challenges and circumstances while working within an economic framework that necessitate choice. It contrasts with developing forces based on a specific threat and scenario" (Davis, 2002, p. xi). Its implementation should "emphasize flexibility, adaptiveness, and robustness of capability. That *implies* a modular, building-block approach" to operational planning, including the following:

- identifying capabilities needs
- assessing capabilities options for effectiveness in stressful building-block missions (i.e., operations)
- making choices about planning targets and ways to achieve them, and doing so in an "integrative portfolio framework" that addresses future warfighting capabilities, force management, risk trade-offs, and related matters in an economic framework (Davis, 2002, p. xi).

In the civilian context, the underpinning for broad, national capabilities-based preparedness (synonymous with *capabilities-based planning*) can be found in the DHS definition, which mirrors that used in the military: "preparing, under uncertainty, to provide capabilities suitable

Figure 3.3
Risk Management Strategies

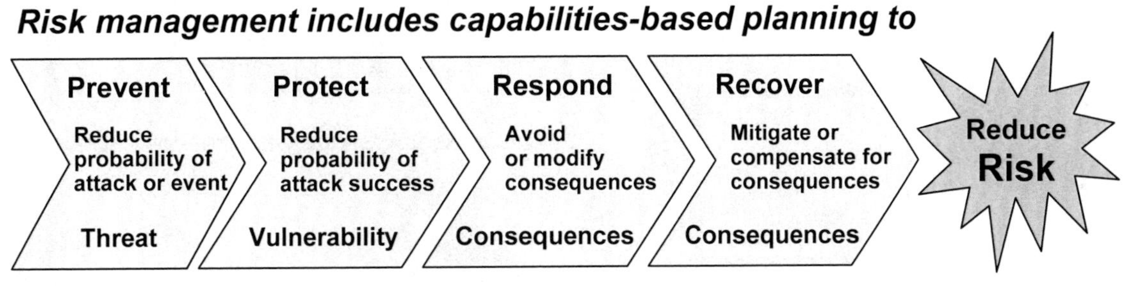

RAND *TR764-3.3*

for a wide range of challenges while working within an economic framework that necessitates prioritization and choice." Further, DHS states that capabilities-based preparedness is "a way to make informed choices about how to manage the risk and reduce the impact posed by potential threats. It focuses decisionmaking on building and maintaining capabilities to prevent and protect against challenges . . . and to respond and recover when events occur" (NPG, p. 30).

Capabilities-based planning includes an iterative and ongoing process to assess current capabilities, determine capabilities gaps, make investment decisions, and reassess capabilities levels. Taken as a whole, the process is dynamic, iterative, and interdependent. The process flows from national-level guidance included in capstone documents published by the White House, DHS, and other interagency members—notably (in the context of this project), DoD, VA, and HHS. The process will be informed by scenarios specific to a community as well as the strategic interests of the community's public and private stakeholders who are engaged in disaster planning. It follows that using a well-designed capabilities-based planning model in a coordinated and collaborative manner will be critical to successful local efforts.

Linking Risk Assessment and Capabilities-Based Planning

Risk assessment and capabilities-based planning are both important, but they are not systematically connected in practice. Because DoD asked RAND to address local risk-informed capabilities-based planning, we sought to conceptually connect the two, based on national policy and described methods for each. This would, in turn, guide our development of a preparedness support tool. Figure 3.4 depicts this connection and is the basis for both the questions we asked during our interviews (Chapter Four) and the presentation of the proposed

Figure 3.4
Process for Applying Risk Assessments for Capabilities-Based Planning

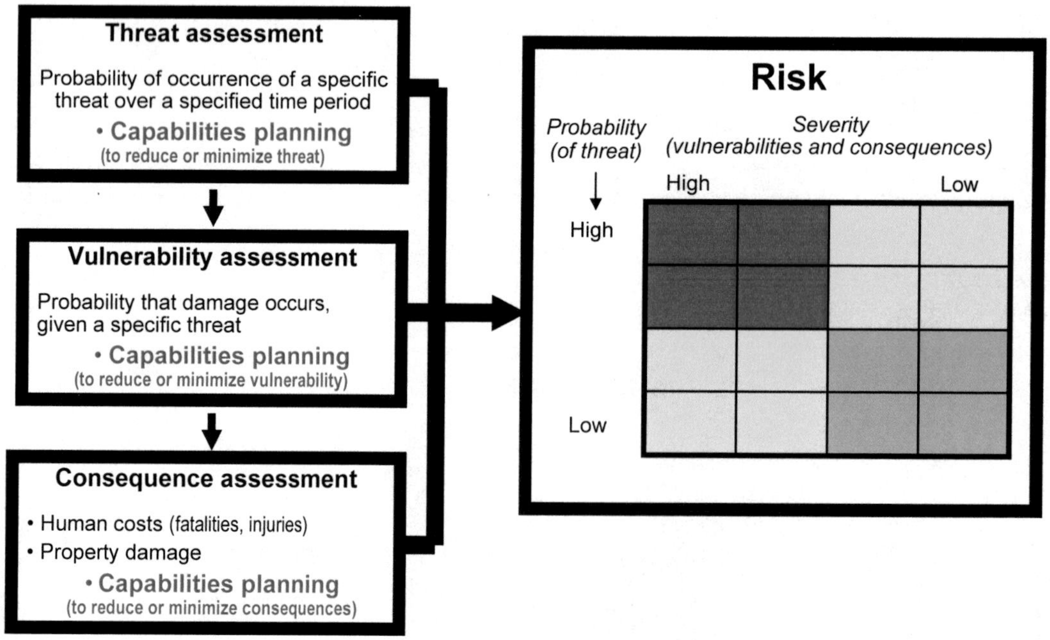

RAND *TR764-3.4*

design for the preparedness support tool (Chapter Six). The left-hand elements of the figure depict the three components of risk assessment, as described earlier, resulting in different levels of risk, as shown on the right-hand side of the figure. Assessment of capabilities to reduce threat, vulnerabilities, and consequences, as shown in red in the figure, introduces the iterative process of risk assessment and planning for capabilities to reduce one or more components of risk.

CHAPTER FOUR

Local Level Civilian and Military Disaster Preparedness Activities

The goal of this project is to create a planning tool that complements effective existing tools, bridges gaps, reflects community needs, and is compatible with current community approaches to disaster planning. To that end, we conducted site visits at five locations to understand better how communities actually engage in disaster preparedness, and to identify the desired features and potential capabilities for a new planning support tool. Our expert interviews with local military installation personnel, civilian actors, and representatives of local VA facilities (see Appendix D) provided insights that helped to shape the framework for the RAND tool, discussed in Chapter Six. In this chapter, we synthesize the findings from across the five site visits. (Details for each site appear in Appendix E.)

Methods

Site Selection

Given the exploratory nature of our research, we chose to use an inductive, insight-generating approach for our site visits. Accordingly, we used purposive sampling (Eisenhardt, 1989) rather than random sampling to select locations. While this approach limits the generalizability of our findings, it was more appropriate than random sampling, given that the purpose of the site visits was to inform the development of a planning tool rather than to test hypotheses pertaining to communities nationwide. We selected locations to fill categories based on the following criteria:

- local military presence (Army, Air Force, or Navy)
- presence or absence of local VA medical centers
- geographic diversity (as indicated by Federal Emergency Management Agency, or FEMA, region)
- population density
- overall population
- size (acreage or square miles)
- primary threats (natural disaster, public health, or terrorism)
- disaster preparedness funding.

We intentionally drew our samples from communities with an active duty military installation, ensuring, as part of that process, that we included at least one for each military department. Although, as noted earlier, our sample was not representative, we did seek variation in other dimensions in order to represent a range of communities that would ultimately use the planning tool developed by this process. For example, we sought locations that differed geographically, as indicated by their FEMA region, which was important, in part, due to potential variation in primary threats: Locations on the eastern seaboard often contend with hurricanes, while those on the West Coast and in the southwestern part of the United States have a higher risk of earthquakes. We also used such measures as population density, overall population, and area size as indicators of the size of the community that needed to be accounted for in disaster preparedness efforts, and we sought variation across this dimension. We also viewed the receipt of disaster preparedness funds as a proxy for a higher-level or more advanced stage of disaster preparedness within a community. Given our goal to learn from communities with disaster preparedness efforts, including military-civilian and civilian-civilian collaboration, that were well under way, we focused on communities that were awarded grants from at least one of three prominent sources of such funding: the Nunn-Lugar-Domenici Domestic Preparedness Program (which no longer exists), CDC's CRI, and DHS's UASI.

We obtained data for each of these measures from published sources (e.g., 2000 U.S. Census, grant-conferring organizations) and developed a list of approximately 15 potential locations for our site visits. We then worked with our research sponsor and military service headquarters staff to identify locations both well suited to our research goals and available to participate. Given the Global War on Terrorism and other military operational demands during 2007, when we were in the process of selecting sites for interviews, it was both challenging and critical to select locations where local military personnel would be available to participate in our study. From our larger set of possible locations, we narrowed our list to five communities:

- San Antonio, Texas, metropolitan area
- Norfolk/Virginia Beach, Virginia, metropolitan area
- city of Columbus and Muscogee County, Georgia
- city of Tacoma and Pierce County, Washington
- city of Las Vegas and Clark County, Nevada.

Table 4.1 lists the military installations that correspond to each location and shows where each community falls in terms of our site selection criteria. For example, the San Antonio area is home to Lackland Air Force Base (AFB), Randolph AFB, and Fort Sam Houston. Its population exceeded 1.1 million people in 2000, and its population density indicates a highly populated urban area. The primary threats to San Antonio include hurricane and pandemic influenza. It was rated as a tier VII city[1] based on the Risk Management Solutions Probabilistic Terrorism Model (Willis, LaTourrette, et al., 2007). The city has received grants from all three sources of funding we considered during the site selection process.

As Table 4.1 demonstrates, we achieved the desired variation on many dimensions: Communities were located in different parts of the country, ranged in community size, faced different types of natural disasters, and were rated differently in terms of the threat of terrorism. Across the locations, we had three sites with a local Army installation (San Antonio, Columbus/Muscogee, and Tacoma/Pierce), three with at least one Air Force base (San Anto-

[1] City tier rankings are based on Risk Management Solutions Probabilistic Terrorism Model, published in Willis, LaTourrette, et al. (2007). The lower the tier number, the higher the terrorism risk.

Table 4.1
Site Visit Selection Criteria

Site							Primary Threat			Funding		
No.	Community (FEMA region)	Installation (service)	VA	Population Density	Population	Size	Natural Disaster[a]	Public Health[a]	Terrorism Tier	Nunn-Lugar-Domenici[b]	CRI[c]	UASI[d]
1	San Antonio, Texas (VI)		OC/VC	2,809	1,144,646	408	Hurricane	Pandemic influenza	VII	Yes	Yes	Yes
		Lackland AFB (USAF)			18,957	2,719						
		Randolph AFB (USAF)			3,560	3,129						
		Ft. Sam Houston (USA)			7,975	3,106						
2	Norfolk, Virginia (III)		VC	4,363	234,403	54	Hurricane	Pandemic influenza	Unranked	Yes	Yes	Yes
	Virginia Beach, Virginia (III)			1,713	425,257	248			Unranked	Yes	Yes	Yes
		Oceana NAS (USN)			4,337	13,390						
		NS Norfolk (USN)			56,065	3,980						
		Little Creek NAB (USN)			7,891	2,373						
3	Columbus, Georgia (IV)		OC	860	185,781	216	Hurricane	Pandemic influenza	Unranked	Yes	No	No
		Ft. Benning (USA)			21,693	171,873						

Table 4.1—Continued

Site							Primary Threat			Funding		
No.	Community (FEMA region)	Installation (service)	VA	Population Density	Population	Size	Natural Disaster[a]	Public Health[a]	Terrorism Tier	Nunn-Lugar-Domenici[b]	CRI[c]	UASI[d]
4	Tacoma, Washington (X)			3,865	193,556	50	Earthquake	Pandemic influenza	Unranked	Yes	No	No
		Ft. Lewis (USA)			22,480	86,042						
		McChord AFB (USAF)			3,710	4,639						
5	Las Vegas, Nevada (IX)		MC/OC/VC	4,223	478,434	113	Earthquake	Pandemic influenza	III	Yes	Yes	Yes
		Nellis AFB (USAF)		8,093	14,161							

SOURCES: Population density: 2000 U.S. Census, people per square mile. Active duty personnel data for military installations: 2006 Force Readiness Manpower Information System (FORMIS) data set. Civilian population data for communities: 2000 U.S. Census. Military installation data in acres from Office of the Deputy Under Secretary of Defense (Installations and Environment) (2005). Community data in square miles from 2000 U.S. Census.

NOTE: OC = outpatient clinic. VC = veterans' clinic. NAS = naval air station. NS = naval station. NAB = Navy amphibious base. MC = medical center. USA = U.S. Army. USAF = U.S. Air Force. USN = U.S. Navy.

[a] Natural disaster and public health threats correspond to National Planning Scenarios.

[b] Communities received different levels of support under Nunn-Lugar-Domenici, but all received an award during at least one funding year.

[c] Indicates whether the community received FY 2006 CRI award.

[d] Indicates whether the community was eligible for FY 2007 UASI award.

nio, Tacoma/Pierce, and Las Vegas/Clark), and one with a strong Navy presence (Norfolk/ Virginia Beach). Three of our locations included multiple military installations, suggesting the possibility not only for local military-civilian collaboration but also for collaboration across installations from the same or sister services. The other two sites had one military installation within their boundaries.

Data Collection

After we identified our site visit locations, the research sponsor worked with each military department to secure the participation of local military personnel and obtain a local point of contact for each installation to aid in site visit planning. Our sponsor also sent letters to the elected leadership of each community (e.g., mayor, county commissioner) to introduce the study, underscore its importance, and encourage their involvement. Lastly, the research sponsor coordinated the participation of local VA representatives. After these initial contacts, members of the RAND project team provided a short introduction to the study and a list of study topics, including representative questions, to intended participants. Recruitment took place from late 2007 through early 2008; the site visits themselves were conducted from April through September 2008; and data-collection efforts ceased in December 2008.

The site visits consisted primarily of expert interviews with civilian stakeholders, local military personnel (both active duty and civilian), and a local VA designee. Our initial focus was on the largest city in close proximity to the local military installation(s); however, at all locations, we included county-level civilian agencies as well, such as county public health departments and sheriffs' offices. If locations had consolidated city-county agencies or governments, we included both city- and county-level professionals in our sample of civilian experts. Additional data-collection efforts involved a social network survey, which was analyzed separately and is the subject of the next chapter.

We developed distinct interview protocols for military personnel and civilian actors, including VA representatives. Later in the study, we developed a separate protocol to obtain information from executive-branch decisionmakers, e.g., mayor's office officials. All protocols appear in Appendix D. We used a semi-structured interview approach: We consistently posed questions related to individual and community background, disaster-related planning and exercises, risk assessment, and needs that could be served by a potential new planning support tool and desired features of such a tool, and we delved into additional, fruitful lines of inquiry as they arose and time permitted. We conducted interviews both on site and via telephone; interviews typically lasted 90 minutes. In total, we conducted 65 interviews with 153 participants. A breakdown of these interviews by community, personnel type (civilian/military/VA), and functional responsibilities is provided in Table 4.2.

Our primary goal was to interview the individuals responsible for emergency management and disaster preparedness; security and law enforcement (including antiterrorism); fire and EMS; public health and medical; and CBRNE or HAZMAT response. During our interviews, we also met with individuals representing other relevant functions, such as civil engineering, public works, public affairs, and executive decisionmaking.[2] However, reflecting the

---

[2] Executive decisionmaker interviews were a late addition to the site visit research design. Even though these interviews were intended to be shorter, they proved very challenging to schedule and were not conducted at all five locations. The insights from the executive interviews we did conduct are reflected in this discussion.

Table 4.2
Site Visit Interview Summary

Site	Overall Number of Participants	Overall Number of Interviews	Number of Interviews, by Function				
			Emergency Management/ Preparedness	Security/Law Enforcement	Fire/EMS	Health/Medical	CBRNE/HAZMAT
San Antonio Metropolitan Area, Texas							
Civilian	5	3	1	1	1	1	0
Military	26	10	4	3	0	5	3
VA	1	1	0	0	0	1	0
Norfolk and Virginia Beach Metropolitan Area, Virginia							
Civilian	8	5	1	1	2	2	2
Military	24	13	5	4	2	4	3
VA	1	1	0	0	0	1	0
City of Columbus and Muscogee County, Georgia							
Civilian	6	4	2	2	1	1	0
Military	14	4	2	3	1	1	2
VA	1	1	0	0	0	1	0
City of Tacoma and Pierce County, Washington							
Civilian	10	5	2	1	1	1	0
Military	31	7	4	3	3	2	6
VA	1	1	0	0	0	1	0

Table 4.2—Continued

Site	Overall Number of Participants	Overall Number of Interviews	Number of Interviews, by Function				
			Emergency Management/ Preparedness	Security/Law Enforcement	Fire/EMS	Health/Medical	CBRNE/HAZMAT
City of Las Vegas and Clark County, Nevada							
Civilian	9	5	2	1	1	2	0
Military	15	4	3	1	1	1	1
VA	1	1	0	0	0	1	0
Total	153	65	26	20	13	25	17

NOTE: EMS = emergency medical services. CBRNE = chemical, biological, radiological, nuclear, and high-yield explosives. Overall number of interviews differs from the number of interviews by function because some interviews included representation from multiple functions. Additionally, in some instances, the same individual represented multiple functions or participated in multiple interviews. Similarly, interviewees categorized as Fire/EMS may also have had HAZMAT responsibilities and capabilities. These dual capabilities are not included in the CBRNE/HAZMAT count.

project scope, we focused on the offices and individuals responsible for overall disaster preparedness and emergency management, risk assessment, and first response during a disaster.

Data Analysis

Analysis from our site visit interviews had two phases: within-case analysis followed by cross-case analysis. Our within-case analysis centered on detailed site visit write-ups intended to be primarily descriptive, lacking researcher impressions and other commentary (Eisenhardt, 1989). To ensure that the site visit write-ups had a parallel structure that would facilitate cross-case analysis, the study team developed a site synthesis guide. This guide, which is provided as Appendix D, provided both detailed instructions about the write-up process and a list of topics that should be covered within the write-up.

Two researchers, one who participated in a specific site visit and one who had not, independently reviewed the interview notes for a specific location and then drafted a list of notes reflecting analysis of patterns and cogent findings for that location. The synthesis guide topics corresponded to those featured in the interview protocols (e.g., definition of community, disaster preparation plans and exercises, preparedness support tool features) and covered background and orientation information for the site, disaster preparedness facilitators and obstacles, and interesting or unique site features.

After two lists were independently drafted for a specific location, the authors reviewed one another's work and consolidated their analyses into one document, resolving conflicting data points and integrating similar list items into single statements. Overall, the independent write-ups were very complementary; few discrepancies were noted. The consolidated write-up was forwarded to a third researcher for review. The third researcher was an individual who had participated in data collection at the location being analyzed but had not participated in the within-case analysis until this point. After all third-party reviews and resultant revisions were completed, the within-case write-ups were used to develop narratives for each location, which comprise Appendix G, and to conduct the cross-case analysis. Specifically, we compared and contrasted sites along each of the dimensions highlighted in the synthesis guide in order to identify patterns across the locations and to understand possible reasons for cross-site variation. Different pairs of locations were analyzed for similarities and differences, with each subsequent pairing refining the emergent findings (Brown and Eisenhardt, 1997). The results of the cross-case analysis are discussed in the remainder of this chapter.

Results

Civilian Community Networks Were Broader Than Local Military Networks

Across the five sites, civilian and military leaders, respectively, articulated fairly consistent definitions of what the community comprised and what constituted their own boundaries for disaster preparedness planning purposes. Civilian interviewees generally had a more expansive view of community that included the main city and county and, in some cases, neighboring counties or districts. Their definitions tended to be bounded by where the population lived and worked. Civilians also viewed the installations as largely independent from the city for purposes of disaster management planning, primarily due to limitations on when military personnel could cross the boundaries of the installation and their perceptions of legal restrictions about aid that that military installation could provide to the surrounding community. At the same time (as will be discussed later), several civilian entities and their military counterparts acknowledged that installations may require assistance from the civilian sector—depending on the nature and scope of a particular incident. Civilians acknowledged that they communicated and worked with military leaders, but they said that planning was not integrated and that, thus, maintaining clear boundaries with the installation was critical.

Military leaders tended to define the community in terms of what they were responsible for in an emergency, which was mostly inside installation boundaries (i.e., "inside the wire"). For example, Nellis AFB personnel identified Nellis and its nonadjacent facilities (Nevada Test and Training Range, Creech AFB) as part of their community because they have direct responsibility for these. Military leaders asserted that their response had to be narrowly focused to an "immediate response" or as first responders in a situation involving a DoD asset on non-federal land (e.g., a military plane crash); such an incident typically involved establishing a National Defense Area (NDA), which provides DoD with temporary authority over the territory in order to safeguard DoD assets.

The VA representatives define the boundaries of their community more similarly to the civilian community—and, in particular, to the public health side of the civilian community—than to the military community. In the Norfolk area, the VA works closely with the hospitals involved in MMRS. The interviewee specified that "MMRS has really helped to bound the community." In the broader Puget Sound region of which Tacoma is a part, the VA works closely with the command hospital in King County, which functions as the Disaster Medical Control Center (DMCC)[3] for the county and the National Disaster Medical System. At a higher level, the VA participates in a FEMA/Emergency Management Strategic Health Care Group, which has a liaison at regional VA headquarters or Veterans Integrated Service Network (VISN).

There was some variation in substance in our four VA interviews for several reasons, the most prominent of which was differences in geographic accessibility. We sought interviews at the VA facility that was closest to the community in which the majority of our interviews took place (i.e., San Antonio, Norfolk, Columbus, Tacoma, or Las Vegas). The closest VA facility was not, however, always geographically proximate to the community in which we were conducting interviews. An example was our interview with the VA representative closest to Columbus, Georgia, who was located several hours away in Alabama and hence did not have any formal ties to Columbus disaster preparedness organizations. At the opposite end of the spectrum, the medical facility in Las Vegas is actually a shared VA/DoD facility. As a result, much of the direct coordination that occurs between VA facilities and civilian public health organizations in other communities is coordinated by DoD personnel at the Mike O'Callaghan Federal Hospital (MOFH).

Despite the general differences in the definition of what constitutes *community* for an installation and for local civilian planning agencies, there were functions in which these distinctions did not strictly apply (e.g., public health and public safety). Civilian public health agencies used a broader definition of *community* because, in most cases, they have to plan to provide services to DoD personnel and dependents, who often live off base in the surrounding community and must be prepared for a public health emergency like any other local resident. For example, in San Antonio, as in other communities, the public health department included

[3] The DMCC is specific to Washington State, as far as we know. One hospital in each county is assigned the role of command hospital, and that hospital takes responsibility for coordinating communications and activities in case of a disaster.

the military population in planning for the SNS, or countermeasures in the event of pandemic influenza or a bioterrorism event. In addition, those communities that had an MMRS organization, such as the Norfolk/Virginia Beach area (Hampton Roads MMRS), considered the broader region, including installations, in their planning for medical response.

Another area in which the community boundaries are often broadened is security. Although military security personnel typically did not generally engage in law enforcement activities outside the installation (due to the provisions of the Posse Comitatus Act and restrictions elsewhere), leaders on both sides reported that intelligence sharing was becoming more common across military and civilian agencies. Coordination at the tactical level was also fairly common in areas where military and civilian communities had to collaborate in order to plan for large-scale annual events.

Military and Civilians Plan Separately but Use an All-Hazards Approach and Often Participate Together in Exercises

Military Installation Planning and Exercising Are Somewhat Piecemeal, with Final Plans Generally an Assembly of Functional "Pieces." Military installations have been instructed to approach major disaster planning from an all-hazards perspective, and this was the case at the locations we visited. Such planning entails examining all potential threats (manmade and natural) collectively, then planning for specific responses using functional and event-specific annexes. Plans are either developed cooperatively via multidisciplinary teams or created in function-specific teams (e.g., antiterrorism and medical) and submitted up the chain of command, where they are integrated into a full plan. When plans are high level, functional organizations may develop their own checklists on how to execute their component of a plan. Typically, there is a top-level version of the emergency response plan that can be used by the emergency operations center (EOC) as a guiding reference, and then there are more detailed plans for each of the emergency support functions (e.g., fire, EMS) that provide more specificity about resources needed and how response should be conducted.

We observed some differences by installation and service in terms of process; however, the end product of an all-hazards plan with functional annexes is comparable across the military. For example, at Lackland AFB in San Antonio, each functional organization on base has its own checklist stemming from the Air Force Comprehensive Emergency Management Plan (CEMP) 10-2 (AFI 10-2501).[4] The first responders (fire, medical, HAZMAT, and the emergency planning team) meet quarterly to discuss the plans. The emergency planning team then synthesizes these checklists and updates the commander quarterly about the development and progress of plans. On the other hand, at the time of our research, Fort Sam Houston did not

[4] According to AFI 10-2501 (p. 46),

> The installation CEMP 10-2 provides comprehensive guidance for emergency response to physical threats resulting from major accidents, natural disasters, conventional attacks, terrorist attack, and CBRN attacks. As such it is intended to be a separate installation plan and will not be combined with other plans until HQ USAF [Headquarters, U.S. Air Force] develops and fields a template and provides implementation guidance. All installations must develop a CEMP 10-2 using the AF [Air Force] template to address the physical threats to their base. . . . The CEMP 10-2 should be coordinated with . . . other installation plans such as the AT [Antiterrorism] Plan 31-101, Base Defense Plan, MCRP [Medical Contingency Response Plan], ESP [Expeditionary Support Plan] and Installation Deployment Plan. The CEMP 10-2 must be coordinated through all tasked agencies and should be coordinated with all units/agencies on the installation. Any conflicts with other plans must be resolved before publication. Readiness and Emergency Management Flights will provide an information copy of the CEMP 10-2, unless it is classified, to local civilian agencies as part of their total coordination effort.

have a global plan like the CEMP 10-2. Rather, installation planning is conducted at the installation headquarters, and all of the tenant units submit their plans to that office.[5] Similarly, Fort Lewis and McChord AFB develop all-hazards plans, but the process is similar to the Army and Air Force processes in San Antonio, respectively, in terms of where central planning occurs and how the plans are shared for review by each functional organization. Planning approaches in the Navy appear to be comparable to those of other services, based on our analysis of the Norfolk region. Planners have access to templates from the Commander, Navy Installations Command (CNIC), which also published Commander, Navy Installations (CNI) Instruction 3440.17, Navy Installation Emergency Management Program Manual. Each installation creates a base plan with functional annexes and event-specific annexes using these common templates and guidance.

In general, risk assessment usually precedes planning and informs the steps of these plans. Plans are fluid and can be modified with data from exercises. Plans are typically drafted and modified via stakeholder input, then further refined following exercises. Thus, exercises are a critical process for ensuring that plans are logistically sound.

Exercises (tabletops, functional drills, and larger-scale field exercises) are used not only to test and refine plans but also to meet external requirements. The frequency of exercises varies, but most installations report conducting them regularly. Field exercises are less frequent for practical reasons, but some installations, such as Fort Benning, indicate that they conduct at least one per year. Nellis AFB conducts more than 50 exercises per year, including tabletops and functional drills across a wide range of issues, some of which relate to disaster preparedness. As we learned during our September 2008 site visit, AFI 10-2501 required a mass casualty exercise at least annually, and AFI 41-106 has a similar requirement for a CBRNE exercise involving off-installation responders.

Typically, each functional team has its own set of exercises, and there are opportunities for integrated military exercises as well. For example, at NAS Oceana and Little Creek NAB in Norfolk, security forces personnel conduct a monthly exercise, and the fire staff does its own training and exercises. However, once per quarter, there is a functionally integrated exercise in which the EOC is operational and all functions (e.g., fire, security) participate. CNIC conducts four integrated exercises per year (the Reliant and Citadel series). Fleet Forces Command conducts the annual Navy-wide Solid Curtain exercise, which tests naval base security forces and other security personnel, and HURREX (a hurricane-related exercise).

Multiple Local Agencies Usually Participate in the Development of Civilian Plans and Exercises. Community planners at the five sites also use an all-hazards framework for disaster planning. Civilian plans generally involve participation of those responsible for emergency management, public safety, fire services, EMS, public health, and hospitals, to ensure that all functions required to respond to an event are included from the early stages of planning. Interaction among civilian agencies often occurs initially in the planning stage via local emergency planning committees.

Overall, the communications between civilian agencies are quite strong, although, in some areas, such as Norfolk, for example, with multiple overlapping local and regional jurisdictions, there is still room for improvement. For example, in Tacoma, there are planning com-

---

[5] In December 2008, eight months after our site visit to the San Antonio area, Army Regulation (AR) 525-27, Army Emergency Management Program, was published. That likely has affected the emergency management planning process employed by Fort Sam Houston personnel and the output produced.

mittees for emergency management, such as its Emergency Preparedness Health Care Coalition. In addition, there are local emergency preparedness coordinating committees, chaired by county health departments, which work on developing integrated plans, and local emergency planning committees often coordinated by county-level officials. In Las Vegas, the interactions across civilian agencies, including the private sector, and with Nellis AFB are facilitated by the Local Emergency Planning Committee (LEPC), chaired by Clark County's Office of Emergency Management. Each agency has specific exercise requirements, which foster more collaboration because agencies are invited to participate in multiagency exercises. In the Norfolk or Hampton Roads, Virginia, area, the leaders in the 16 districts coordinate via the Hampton Roads Emergency Management Committee and the Hampton Roads Regional Emergency Management Technical Advisory Committee.

In general, planning entails an annual process of drafting or revising plans, including additions or changes to expected capabilities and contact information. Civilian planners rely on several tools to guide the development of plans. These include the NRF, the TCL, other FEMA guidance, the Homeland Security Exercise and Evaluation Program (HSEEP); recommendations from CDC (primarily for public health preparedness), and state emergency planning templates. For example, planners in San Antonio and Las Vegas report using information from Texas, given the comprehensiveness and accessibility of the Texas templates. Texas has a template for basic emergency response plans and 22 functional annexes. San Antonio uses this template to develop standard operating guidance, then partners with the Texas Engineering Extension Service at Texas A&M University for technical assistance. All 22 annexes are submitted through a coordination process at the weekly emergency planning committee meetings for review of consistency and clarity about functional expectations. This process also allows for planning around grant dollars to ensure that funding is appropriated correctly and that agencies are not working at cross-purposes to meet grant requirements.

Like their military planning counterparts, civilian leaders organize several exercises per year, including tabletops, functional drills, and larger-scale field exercises. Each agency has specific requirements for exercises. For example, public health departments engage in pandemic influenza and SNS exercises. In Columbus, planners develop a standard mass casualty exercise and other exercises specific to local events, such as a dam break or a workplace hostage scenario. In the Norfolk/Virginia Beach area, the MMRS indicates that much of its grant guidance focuses on selecting projects in high risk areas, based on the National Planning Scenarios. The guidance dictates how they exercise. For example, in recent months, the MMRS has focused on the threats of multiple improvised explosive devices (IEDs), aerosolized anthrax, major hurricanes, and cyber attacks.

Joint Military-Civilian Planning and Exercising Mostly Takes Place Once Independent Plans Have Been Completed. The cross-site analysis reveals one common pattern of military-civilian interaction in planning and exercising for major disaster response: Typically, military and civilian leaders create their plans initially in isolation without input from the other entity in the development of those plans. However, once the plans are prepared, the level of dissemination and collaboration varies, ranging from sharing the plans to participation in joint meetings and exercises. Much of this variation can be explained by recent experiences, including emergencies and deployments, functional need, and geography, particularly in the case of the VA and National Guard.

The extent of military and civilian participation in planning meetings and exercises varies across sites. For example, in Norfolk, military-civilian interaction is at a nascent stage. Collaboration is increasingly recognized as important, but, at the time of our site visit (May 2008), there appeared to be relatively little integrative planning and few exercises. In San Antonio, the military has become more involved with the civilian side of emergency management, especially since Hurricanes Rita, Gustav, and Ike; military representatives have become part of the community EOC and are at the table for exercises. Despite this change, the military is still typically not included in community plans, in part because it is not clear what capabilities it will be able to contribute given fluctuations in the military's local capabilities (due to operations tempo, or OPTEMPO) and its focus on internal installation matters (especially for events involving terrorism). In Tacoma, military and civilian organizations have a relatively high level of interaction with respect to emergency planning and exercises, but the bases are not engaged in joint planning with the local civilian community. Despite several DoD requirements for inclusion of civilian counterparts in developing their disaster preparedness plans, the military installations we visited do not involve civilian organizations in *developing* installation-level all-hazards response plans. Military installations may share plans with civilian counterparts as a courtesy and verify that the contact information for community organizations is correct. In Columbus, Fort Benning leaders are invited to exercise with civilian leaders quarterly, at least in conducting tabletop exercises, but Columbus leaders are typically less involved in Fort Benning planning and exercises, with the notable exception of those related to the annual Western Hemisphere Institute for Security Cooperation (formerly, the School of the Americas) protest.

Although there is discrepancy in relative participation, the level of interaction between military and civilian leaders generally depends, for planning purposes, on the type or magnitude of event and the function of the specific agency. Local mutual aid agreements (MAAs) tend to be first-responder or functionally focused (fire services, law enforcement, medical), and, therefore, most military-civilian interaction occurs across the same function. Fire services and public health leaders tend to be more connected across military-civilian boundaries, sometimes motivated by a lack of resources. For example, a military fire department typically has less ambulance support and thus may need to call on civilian assistance for emergency medical services. The Public Health Emergency Officer (PHEO) at McChord AFB serves on the Pierce County committee that meets quarterly to develop plans for pandemic influenza. In San Antonio, the public health department works directly with the military bases to develop SNS plans because many DoD personnel and dependents live off base and must be accounted for in enumeration for mass prophylaxis. Military and civilian security forces tend to be less engaged than fire and health, although they do cooperate on a tactical level, particularly during preparations for large-scale annual events, such as air shows.

Las Vegas provides a useful example of how the level of engagement can vary by agency function and event type. Nellis AFB is part of Clark County's LEPC, and Nellis AFB has presence at the Clark County EOC. These connections facilitate greater communications between emergency planners on base and in the local civilian community. Nellis AFB and Clark County share their emergency operations plans; however, the degree to which specific functional organizations within each entity are aware of the counterpart plan is inconsistent. Moreover, there is no civilian involvement in writing parts of the plans for Nellis AFB, though civilians are engaged when there is a need to draft a memorandum of understanding or agreement with a specific Nellis organization (e.g., fire). Although most planning is military or civilian specific, there are large scheduled events like Aviation Nation (the annual Nellis open house and air show), the National Basketball Association (NBA) All-Star Game, or NASCAR

(National Association for Stock Car Auto Racing) races that may involve more coordinated planning because the size and level of support needed overwhelms one entity.

Both Military and Civilian Planners Tend to Interact with Local VA Counterparts in Disaster Preparedness Planning and Exercising. The interviews indicate fairly regular interaction between civilian planners—particularly in the public health arena—and the local VA facilities. For example, in Las Vegas, the MOFH at Nellis AFB is a joint VA/DoD facility. Its link to the civilian community is through an organized group of health safety officers, of which both the MOFH and the Clark County Department of Public Health are members. In Columbus, the closest VA facility, the Central Alabama Veterans Health Care System (CAVHCS), which is responsible for oversight of the Columbus outpatient clinic, is two hours away in Alabama; thus, lack of proximity limits the potential CAVHCS-Columbus partnership. However, the CAVHCS representative is directly involved in the local emergency management committees in both of the counties to which it is geographically proximate: Montgomery and Macon counties, Alabama. In these counties, CAVHCS participates regularly in discussions, training, and local exercises.

Proximity is also an issue in the Tacoma area, where the VA is not a regular stakeholder in planning and exercising because the main office is 35 miles away—in Seattle—from Tacoma's main urban center. As mentioned previously, however, VA Puget Sound does work with the command hospital in King County as part of the larger National Disaster Medical System (NDMS) in the region. In the two regions where there was geographic proximity between the VA and the local community in which we focused our interviews, there was a significant amount of coordination across planning and exercises. In the Hampton Roads area, the VA reported regular drills with the civilian public health community, approximately every three months. The San Antonio VA reported participating in an annual citywide drill and noted that it participates in everything to which it is invited, as doing so fulfills its requirements for Joint Commission accreditation. In San Antonio, the VA also reported exercising NDMS responsibilities and evacuation annually with all federal civilian components in the region. The Puget Sound region reports similar NDMS exercises but only once every three years.

Although not considered to be part of emergency management–related activities, most VA representatives mentioned the VA/DoD Contingency Hospital System Plan when asked about their relationship with local military installations. The contingency plan is the implementation of the May 1982 VA/DoD Health Resources Sharing Act (Pub. L. 97-174), which gave the VA the mission to augment DoD's medical capabilities in the event of war or national emergency. Beyond this statutory relationship, regular interaction between local installations and VA facilities again varies widely based on geographic proximity as well as other factors. In Hampton Roads, the VA has exercised the VA/DoD Contingency Plan, calculating how long it will take to get to and from Langley AFB or the naval air station. Beyond that, there is minimal regular interaction with either of the installations for emergency preparedness activities, although the representative noted that, because the VA hospital is part of the broader public health system in the region, it would be a health-care provider to the military in the same way that any other hospital in the region would provide additional medical capabilities in case of a major disaster. In Las Vegas, the VA representative noted that the VA's leased clinics would likely be closed in the event of a disaster and that all medical personnel would report to the MOFH to backfill. In San Antonio, the VA reported non-participation in installation exercises, which its representative attributed primarily to not having been invited.

In terms of planning, it appears that most of the VA planning documents are separately prepared, although there is some sharing of completed plans after the fact with both the civilian or military sectors. All VA facilities base their plans on the VA's *Emergency Management Program Guidebook* (VA, 2002). In the Norfolk area, the VA uses the VA-specific guidance but also refers to community plans to ensure that they are appropriately tied in. In Las Vegas, although there is no specific sharing of plans with the local community, the VA does send copies of its plan to Nellis AFB for coordination and receives copies of the MOFH's emergency response plan, in which the VA is explicitly mentioned. In Puget Sound, by contrast, the VA regularly shares its emergency preparedness plan with the county but not with local installations. The Puget Sound VA regularly receives a copy of the county plans and has even helped write portions of the plan as part of the representative's involvement in the King County LEPC.

In Contrast, Local Military and Civilian Planners Interact Very Little with the National Guard in Disaster Preparedness Planning. In the five sites, civilian planners have little interaction with the National Guard in planning for disasters—most National Guard planning for disaster assistance is conducted at the state level. There is, however, regular use of the Guard's CST units for responding to specific emergencies. For example, in Las Vegas, civilian planners use the CST during events, such as the annual influx of tourists for New Year's Eve festivities on the Las Vegas strip. The CST works with law enforcement to enhance security efforts, particularly those related to a potential chemical, biological, or radiological event. In Columbus, emergency planners do not plan, exercise, or regularly interact with the CST. However, the civilian community is aware of the Guard's capabilities and considers the CST as a resource particularly for CBRNE-related help during a disaster. In the Tacoma area, Camp Murray—located adjacent to McChord—is a core location for Guard personnel, and there is some interaction between the two. Guard elements are regularly involved in various planning and exercise activities. In addition, the Guard's Western Air Defense Sector (responsible for air defense west of the Mississippi River) is a tenant unit on McChord AFB. Even though there are still official channels to go through to request Guard support, informal relationships with Guard personnel have developed due to the close proximity.

Military and Civilian Planners Carry Out Risk Assessments and Capabilities-Based Planning Based on Different Tools Available to Them

Overview. Figure 4.1 serves as a reminder of the way we view, for purposes of this project, the risk assessment process and its relationship to capabilities-based planning.

Summarized in this section are highlights of our findings about the risk assessment process at the five sites, followed by more detailed descriptions of the various components of risk assessment—threat, vulnerability, and consequences.

Military. Risk assessment for a military installation is a broadly standardized process in the sense that all installations are required to meet established DoD benchmarks for antiterrorism protection (DoDD 2000.12). The process quickly becomes unique, however, as one examines the varied configurations of players across the different installations, locations, and services.

A risk assessment begins, in its simplest form, with a threat assessment that identifies the most likely near-term threats—both manmade and natural hazards. This assessment is based partly on indications and warnings from on-site intelligence analysts; historical data about natural disasters to which the geographic location is most susceptible; and information

Figure 4.1
The Risk Assessment Process and Its Relationship to Capabilities-Based Planning as Used in This Report

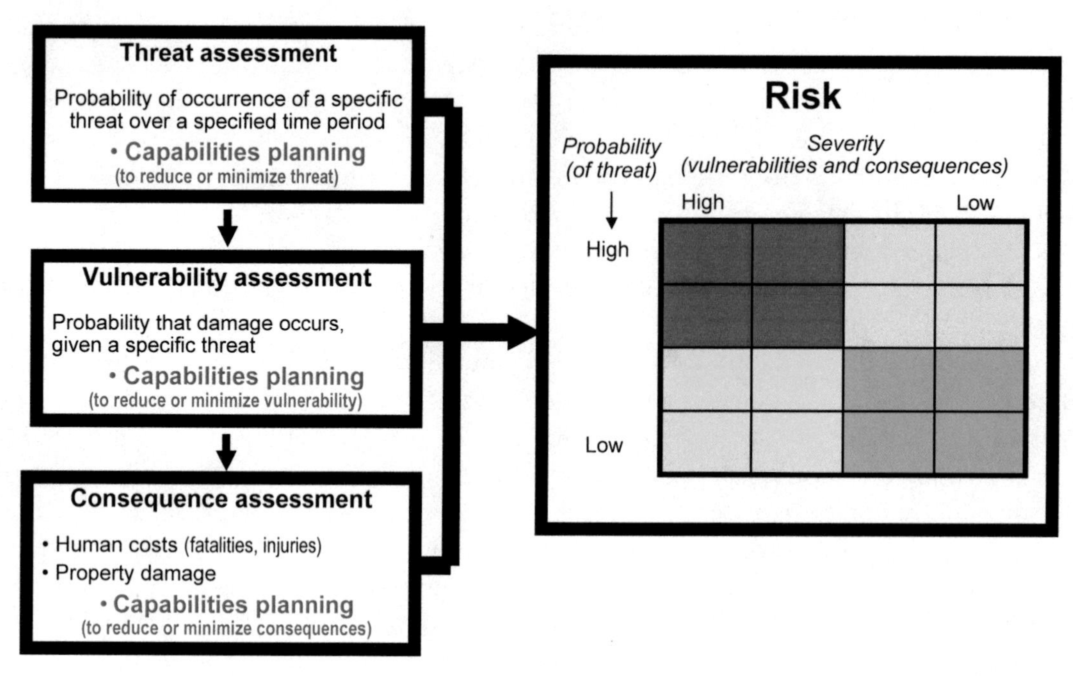

RAND *TR764-4.1*

obtained from Fusion Centers or shared by local police, FBI, or military criminal investigation organizations. There is some sharing of threat assessment results with the civilian community upon the threat assessment completion, but most threat assessments seem to be completed separately and kept from broad circulation. Threat working groups on most installations meet quarterly and enable the various participants to come together regularly and share information; in some cases, similar civilian-led threat working groups are increasingly conducting comparable activities.

Typically, when the threat assessment has been completed,[6] the installation commander or his or her designee conducts a vulnerability assessment to match potential threats with the most vulnerable and most attractive targets and calculate potential consequences.

The next step is to identify and develop a mitigation plan; implement the identified strategies in order to offset the negative effects of potential attacks; and assess the effectiveness of the mitigation strategies (e.g., by conducting tabletop or other exercises). DoD guidelines currently require a higher-headquarters antiterrorism/force protection assessment of each DoD installation every three years (DTRA, undated). To support that requirement, DoD has developed or adapted a number of methodologies, frameworks, and tools to help users conduct vulnerability assessments. The most commonly reported of these is the Joint Staff Integrated Vulnerability Assessment (JSIVA) program run by the Defense Threat Reduction Agency (JP 3-07.2, p. B-2). Installations generally complete the threat/vulnerability/mitigation risk assessment process annually.

---

[6] At times, interviewees seemed to imply that the vulnerability assessment could either precede or follow the threat assessment.

Civilian. Although the broad risk assessment processes are similar in structure in the military and civilian communities, the civilian threat assessment process is somewhat more ad hoc than the military one. The risk assessment process overall is looser and less standardized both in terms of having access to definitive guidance from above and in terms of how the many players in the civilian emergency management community are expected to fit seamlessly together. Often, the threat assessment piece of the process is somewhat isolated from the other pieces. This is likely due in part to the sensitive nature of threat information and lack of clearances on the part of civilian emergency planners and in part to the decentralized nature of the civilian community. The civilian community does conduct its own assessments through a hazard vulnerability assessment process, as exemplified by the Hazard Vulnerability Analysis assessment tool created by Kaiser Permanente (see Figure 4.2, which shows an example for natural hazards). This allows the user to derive a relative threat assessment based on a calculation of the probability of various threatening events occurring, the impact should the event occur, and the capabilities that would be required to meet that threat. The threat/event list is generally pre-populated across three categories: natural, technological, and manmade.

Department of Veterans Affairs. Emergency managers with the VA have processes similar to those in the military community in terms of the top-down direction of, and requirement to conduct, a risk assessment. Each step of the risk assessment and subsequent exercise and planning process for emergency management is related in detail in the VA's 2002 Emergency Management Program Guidebook, which includes a hazard vulnerability analysis tool (VA, 2002), which is very similar to the Kaiser Permanente Hazard Vulnerability Analysis assessment tool just described. VA emergency managers are more similar to their civilian counterparts when it comes to conducting threat assessments, however, in that they do not have access to on-site intelligence analysis for manmade threats and tend to focus on historical experience and common sense. At least one of our interviewees specified a close relationship with local law enforcement, which, in theory, would provide the VA with a sense of the scope of the local malicious threat.

Local Military Threat Assessments Use Highly Variable Processes

A comparison of our five sites suggests that the threat assessment process is the most variable part of the larger risk assessment because different functional offices are involved in different pieces of the broader threat assessment depending on the service, installation, and functional area of responsibility. At the most extreme, CBRNE; toxic or industrial chemicals and toxic or industrial materials; public health threats to water, power, and food; as well as natural disaster threats are each assessed separately from terrorism, and both civilian and military representatives indicate significant reliance on statewide assessments of natural disaster threats, such as volcanoes or tsunamis.

Our site visits provide a number of examples of both differences in the processes used to conduct assessments and the stakeholders whose input is sought. At Fort Sam Houston, there is no single, integrated threat assessment process. The installation antiterrorism officer (ATO) assesses all-source intelligence for the base's annual threat assessment. The ATO is specifically responsible for assessing terrorism-related threats, an assessment that the officer conducts with input from military intelligence, local law enforcement, the local FBI office, and the Joint Terrorism Task Force representatives. Non-terrorism threats that do not fall within the antiterrorism officer's realm are assessed by functional communities on the installation. The PHEO assesses public health threats, such as pandemic influenza; the CBRNE

Figure 4.2
Assessment of Naturally Occurring Events, from Kaiser Permanente Hazard Vulnerability Analysis Assessment Tool

Hazard and vulnerability assessment tool
Naturally occurring events

KAISER PERMANENTE®

Event	Probability	Severity = (magnitude – mitigation)						Risk
		Human impact	Property impact	Business impact	Preparedness	Internal response	External response	
	Likelihood this will occur	*Possibility of death or injury*	*Physical losses and damages*	*Interruption of services*	*Preplanning*	*Time, effectiveness, resources*	*Community/ Mutual Aid staff and supplies*	*Relative threat**
Score	*0 = n.a.* *1 = Low* *2 = Moderate* *3 = High*	*0 = n.a.* *1 = Low* *2 = Moderate* *3 = High*	*0 = n.a.* *1 = Low* *2 = Moderate* *3 = High*	*0 = n.a.* *1 = Low* *2 = Moderate* *3 = High*	*0 = n.a.* *1 = High* *2 = Moderate* *3 = Low or none*	*0 = n.a.* *1 = High* *2 = Moderate* *3 = Low or none*	*0 = n.a.* *1 = High* *2 = Moderate* *3 = Low or none*	*0 - 100%*
Hurricane								0
Tornado								0
Severe thunderstorm								0
Snow fall								0
Blizzard								0
Ice storm								0
Earthquake								0
Tidal wave								0
Temperature extremes								0
Drought								0
Flood, external								0
Wild fire								0
Landslide								0
Dam inundation								0
Volcano								0
Epidemic								0
Average score	0.00	0.00	0.00	0.00	0.00	0.00	0.00	0

**Threat increases with percentage.*

Risk	= probability	× severity
0.00	0.00	0.00

NOTE: n.a. = not applicable.

RAND *TR764-4.2*

officer assesses CBRNE-related threats, such as those involving the railroad and potential toxic chemical accidents. Brooke Army Medical Center, for example, specified that its primary role in terms of threat assessment is to assist both the military and civilian communities with disease surveillance.

At Randolph and Lackland AFBs, the Air Force Office of Special Investigations (OSI) conducts threat assessments, as required by AFI 10-245. This assessment includes both criminal and terrorist activity, and results in a classified threat ranking (based on likelihood and feasibility), which is maintained by the installation ATO. Core members of the terrorism working group, including the ATO and installation security forces, are involved in the threat assessment process for the base. The larger working group meets every other month and is responsible for conducting other assessments—from the possibility of pandemic influenza to toxic industrial chemical accidents throughout the San Antonio region. Threat information about infectious diseases comes from CDC, military hospitals involved in disease surveillance, and the civilian public health community.

The process employed at NS Norfolk is a mix of the approaches we found in the other sites. Like Randolph and Lackland AFBs, Norfolk relies on a service-level investigative unit, the NCIS, which is the U.S. Navy's primary investigative unit responsible for "actual, potential or suspected terrorism, sabotage, espionage, and subversive activities" (NCIS, undated). Installation security forces assess risks for certain events, including regularly scheduled major events on the base. Naval installation security forces are in close contact with NCIS, meeting with local field officers at least once per month and talking to them once or twice per week. NCIS is the more direct interface between the Navy, the FBI, and local police departments. However, just as Fort Sam Houston relies on functional experts to assess non-terrorism threats, so too does Naval Medical Center (NMC) Portsmouth. We learned that the medical center's internal threat assessment is conducted partly by subject matter experts who assess threats for their particular areas of responsibility and expertise. Those assessments are then combined with information about current events from such sources as CDC and DHS. One interviewee noted, "If the CDC says something is big, then we start throwing assets at it."

One commonality that spans installations, regardless of the military service, and locations, is that the ATO—working closely with base security forces—takes the lead in the terrorism piece of the threat assessment process. The ATO's specific emphasis is on terrorism-related threats to the area within the boundaries of the installation and to military personnel on official business in the local community (e.g., during a special event). The ATO coordinates with military intelligence, service-specific criminal investigation organizations, and state and local police. Norfolk-area military interviewees noted that there is no specific guidance about where threat information should be derived, and, accordingly, they obtain threat-related information from a variety of different sources, including U.S. Navy and DoD websites, the Virginia Fusion Center and the Virginia EOC (both located in Richmond, Virginia), local and national news sources, review of incoming intelligence, and information received from NCIS's Multiple Threat Alert Center, which provides indications and warning for a wide range of threats to Navy and Marine Corps personnel and assets around the world. NS Norfolk also coordinates with the U.S. Coast Guard and, in particular, with the Coast Guard's Joint Harbor Operations Center (JHOC), which is located on NS Norfolk.[7]

---

[7] The JHOC was established after 9/11 as a prototype for bringing the Navy and the Coast Guard together in a joint operation in order to ensure port security—specifically, monitoring military and civilian vessels entering and exiting the

Local Civilian Threat Assessments Are Even More Variable Than Those by the Military

As mentioned previously, the civilian threat assessment process is more ad hoc than the military process. This is likely due in part to the sensitive nature of threat information and in part to the decentralized nature of the civilian community, as well as the lack of a single source of guidance that both requires and instructs civilian organizations about such assessments. Some civilian emergency managers emphasized their reliance on a common-sense threat assessment based partly on historical events—a perspective that they also extended to their subsequent evaluation of vulnerabilities. Columbus's emergency manager describes the framework he uses as one in which threats are a function of possibility, probability, and consequences and are scored in accordance with "expert judgment." This is essentially the Kaiser Permanente Hazard Vulnerability Analysis assessment tool depicted earlier in Figure 4.2. Some communities have access to an intelligence Fusion Center where information sharing across traditional bureaucratic stovepipes is encouraged (GAO, 2007)[8] or to a Joint Terrorism Task Force (JTTF) (FBI, 2004),[9] which is focused on sharing terrorism-related threat information at the local level. At our sites, however, it was not always entirely clear whether the threat information—which tends to be limited to civilian participants with some type of security clearance—was funneled directly into the broader emergency preparedness planning process. It varied by site.

Civilian interviewees frequently confused the idea of a threat assessment with a vulnerability assessment. Many referred to their hazard vulnerability assessment process when asked about their threat assessment process. Some interviewees spoke specifically to the issue of threats, however. In Columbus, for example, the Muscogee County Sheriff's Office (MCSO) helps both businesses and local government offices to conduct threat assessments. MCSO works with the JTTF in Atlanta, which sends the office monthly intelligence reports in return for any relevant threat information from Muscogee County (for example, reporting that a gas tanker was stolen locally). The JTTF in Atlanta also provides various types of assistance on request, such as Arabic interpreters, and is colocated with Georgia's Information Sharing and Analysis Center—a DHS-sponsored intelligence Fusion Center. An MCSO representative also sits on Fort Benning's Terrorism Working Group.

---

lower part of the Chesapeake Bay. The Coast Guard focuses on threats outside of restricted Navy waters. Another JHOC is currently under development in San Diego, California.

[8] This October 2007 GAO report identified Fusion Centers in at least the planning stages in every state in the nation. The state of California alone had four regional Fusion Centers. Almost all Fusion Centers are led by state or local law enforcement entities and have federal personnel assigned to them. In general, Fusion Centers are mechanisms for information sharing among state, local, and federal entities (e.g., DHS, FBI, Drug Enforcement Administration [DEA], Bureau of Alcohol, Tobacco, Firearms and Explosives [ATF]), as well as collaborative operational efforts "to detect, prevent, investigate, and respond to criminal and terrorist activity" (p. 1). Some Fusion Centers also include personnel from public health, social services, public safety, and public works organizations.

[9] There are more than 100 JTTFs today, including at least one in each of the FBI's 56 field offices, and another 50 or so spread out among other major cities. This includes some 2,196 FBI special agents, 838 state and local law enforcement officers, and 689 professionals from other government agencies (e.g., DHS, Central Intelligence Agency [CIA], Transportation Security Administration [TSA]). JTTFs are operational entities that undertake surveillance, source development, and investigative activities, but also focus on information sharing with local law enforcement. JTTF personnel on the FBI side are often located within the FBI's field and regional offices, and their primary focus is addressing terrorism threats and preventing terrorism incidents. The JTTFs share classified and unclassified information with their federal, state, and local partners and hold meetings for their members and agency liaisons. The regional JTTFs are coordinated at the national level by a centralized JTTF at FBI headquarters within the Strategic Information and Operations Center.

The San Antonio Police Department reports coordinating on a weekly basis with the FBI and some 75 to 80 additional people in order to assess the threat—presumably in the context of the FBI JTTF located in San Antonio. Norfolk is the headquarters for one of the FBI's two Virginia-based JTTFs. The Pierce County, Washington, Emergency Management Office serves as the nexus for that county's Terrorism Early Warning Group (essentially a Fusion Center), which promotes information and intelligence sharing across the civilian, military, and business communities, and the JTTF headquartered in Seattle reports regularly to the various regional coordinating councils in Kitsap, King, and Pierce counties. The state of Washington conducted a hazard vulnerability assessment for the entire state, and each county conducted a county-level assessment. But, at the local level, most emergency management personnel seem to rely on the higher-level assessments, and many of them, such as the Tacoma Fire Department, participate in the Pierce County Terrorism Early Warning Group.

The Department of Veterans Affairs Uses a Standardized Approach to Combined Threat and Vulnerability Assessment

Most of the VA representatives reported basing their threat assessment processes on a combination of historic events and other inputs specific to their local areas, such as the location and accessibility of rail lines, chlorine tankers, and major transportation routes. The VA representative subsequently evaluates the likelihood of the various threats occurring and sends the initial assessment to the Emergency Management Committee (whose membership is specified in the VA EMP Guidebook) for review. The list of threats is also taken from the hazard vulnerability assessment list provided in the guidebook. This tendency to combine vulnerability or opportunity with threat is the biggest difference between the deliverable produced as part of a military threat assessment and that produced as a result of a civilian threat assessment. The Norfolk VA representative spoke directly to this issue, noting that the VA does not have a formal threat analysis, although it draws from whatever the civilian community has produced. Seattle-area VA representatives also stressed their good relationship with the local police and sheriffs' departments in King and Pierce counties—thereby mitigating some of the challenges of accessing real-time threat information.

Military Installations Are Also Better at Vulnerability Assessments

As is the case with the threat assessment process, military installations across the services recognize a broadly similar top-level structure for conducting vulnerability assessments. DoD guidelines currently require a higher-headquarters antiterrorism/force protection assessment of each DoD installation every three years (DTRA, undated). In addition, new post-9/11 emergency response plan requirements mean that installations have had to incorporate planning for natural disasters and manmade accidents into what was previously a limited antiterrorism/force protection plan. Service-based variations in vulnerability assessment processes were similar to those we found for the threat assessment process. The ATO at Fort Lewis uses the JSIVA protocols to begin the vulnerability-assessment process. Fort Lewis also forms functional expert tiger teams,[10] which conduct vulnerability assessments on food and related public health issues. In contrast, the Air Force splits its vulnerability assessments into pieces that parallel the missions of certain functional units and priorities. Bioenvironmental engineering is responsible for

---

[10] In this context, the term *tiger team*, sometimes called a *red team*, normally refers to a group from outside an organization that is engaged to test defenses and other aspects of an organization's preparedness.

assessments of the water supply, while public health assesses the food supply, and medical facilities conduct still other assessments. Public works and myriad other units also conduct critical infrastructure assessments both on and off base. Often, threat and vulnerability assessments happen in conjunction with each other as each building assesses its vulnerabilities and reports back to the ATO, who combines the results of each of these independent assessments.

All military hospitals are subject to the same accreditation standards as civilian hospitals. Civilian agencies have developed a number of frameworks and tools to address their own need for a risk analysis template. One of these is the Kaiser Hazard Vulnerability Analysis assessment tool mentioned and depicted previously—essentially, an Excel spreadsheet with simple, additive formulas, which has been incorporated into many larger emergency management frameworks, including that of the VA (Kaiser Permanente, 2001). Medical personnel we interviewed at each of the five sites we visited were using a hazard vulnerability analysis tool that was similar in structure to the Kaiser tool—probably due to the requirement established by the Joint Commission on the Accreditation of Healthcare Organizations to produce such a hazard vulnerability analysis. At Madigan Army Medical Center, both threat and vulnerability assessments are assessed by the Disaster Preparedness Committee, the latter using the Kaiser Hazard Vulnerability Analysis assessment tool. Each department at Madigan, and each clinic, provides input through a hazard vulnerability assessment, including the PHEO in terms of infectious diseases. This structure was similar at NMC Portsmouth, Virginia.

Local Civilian Planners Use a Variety of Vulnerability-Assessment Methods

Vulnerability assessments are mandated on the civilian side mostly through federal grant requirements, although sometimes by the state. The state-mandated hazard identification and vulnerability analysis for the Pierce County area (which includes the city of Tacoma) is conducted by the Pierce County Department of Emergency Management in collaboration with the Tacoma police. Natural and technological hazards are listed and detailed in the state guidance, and the resource also includes worksheets to facilitate hazard identification and risk assessment, vulnerability analysis, mitigation measures, and mitigation planning. The vulnerability-analysis worksheet breaks down assets into types (e.g., residential, commercial, hospitals, schools, hazardous facilities) and prompts the analyst to input information on the number of people, number of buildings, and approximate value. As business-continuity planning becomes more popular, communities are beginning to see the involvement of large companies in local risk assessment and disaster preparedness planning. Private industry has been particularly involved in efforts to identify and catalog detailed infrastructure. In the case of Pierce County, industry works closely with the Terrorism Early Warning Group; in Columbus, Georgia, industry works closely with the Emergency Management Division of Columbus Fire and Emergency Medical Services.

The Department of Veterans Affairs Guidebook Specifies a Standardized Approach to Vulnerability Assessment

At each of the five sites, the local VA facility uses the Hazard Vulnerability Analysis methods provided in the VA guidebook (VA, 2002) to conduct assessments of VA-owned or contracted facilities in that region. For each organization, the VA calculates how the identified hazards will affect human life, safety, property, and operations as a health-care unit. Each hazard is given a score from 1 to 3 (with 3 indicating either the highest impact or the highest likelihood of occurrence). After totaling each column, each VA planner specifies his or her medical center's standard operating procedures (SOPs) or develops an annex to address any hazard or vulnerability that scored a 2 or higher in any single category, or a total of 6 overall. The initial assessment then goes to that center's emergency management committee (whose membership is specified in the guidebook) for review and final approval.

In the Norfolk area (Hampton Roads), the VA process for establishing vulnerabilities related to the VA/DoD Contingency Plan involves timing the flights into and out of Norfolk NAS and from there to the Hampton Roads VA facility. The VA assessed vulnerability along the path between the various airstrips that would be receiving patients and the roads that would be transporting patients from the airstrips to the VA facilities. This initial assessment also went to the emergency management committee for review and final approval.

We Identified Facilitators and Barriers to Local Disaster Preparedness That Are Important to Consider in Designing a Planning Support Tool

During our expert interviews, we encouraged participants to discuss both the facilitators of local disaster preparedness planning and the challenges or barriers they faced in those efforts. We not only analyzed our interview notes for facilitators and barriers that were salient to interview participants but also identified issues raised during interviews that the study team inferred as either promoting or hindering effective local disaster preparedness. Strategies undertaken to avoid or surmount challenges were noted as well. In this section, we describe facilitators of local disaster preparedness, barriers encountered by military and civilian stakeholders, and strategies employed to limit their impact.

Facilitators.

Grants. Receipt of external disaster preparedness funds, primarily grants, was an important criterion in our site selection process. Each location we visited benefited from some type of external funding; all five received financial support from the Nunn-Lugar-Domenici program, for example. DHS funding for the Fusion Center in Las Vegas is another example of the agency's support to localities across the country, although such funding support was not among our specific site selection criteria.

These funding streams enabled cities to make investments in equipment and training that they likely would not be able to make otherwise. Indeed, several military personnel observed during interviews that local installations lacked the funding the civilian side had to invest in communications-oriented equipment, whereas civilian interviewees in such locations as San Antonio and Columbus readily acknowledged the improvements they were able to make due to grants. In Columbus, they also discussed how grants enabled them to train personnel for a newly created search-and-rescue team.

Grants facilitate disaster preparedness by providing financial support. However, they also provide benefits in terms of external attention, common guidance, and regular interaction, as discussed in the subsections that follow.

External Attention. Attention from external stakeholders, including federal authorities, the media, and the general public, also appeared to spur disaster preparedness efforts. Grants received by the locations we studied were also a type of attention that had a Hawthorne experiment–like effect on the communities;[11] being awarded funds and publicly identified as a grant recipient was a source of pride that seemed to inspire increased collaboration and coor-

[11] The Hawthorne experiments were a series of industrial/occupational psychology experiments conducted in the 1920s and 1930s for the purpose of improving worker productivity and team output. An unexpected finding of the research was

dination. For example, one civilian interviewee in Columbus noted during his interview that Columbus was one of 53 cities designated as a Nunn-Lugar-Domenici city prior to the events of September 2001 and that, although this "raised some eyebrows," it created momentum that propelled the city's disaster preparedness work for many years.

Funding awards were not the only source of external attention about which we learned during interviews. Experiencing a major event also led different stakeholders to examine closely a community's disaster preparedness, both favorably and unfavorably. This was the case in San Antonio after Hurricane Katrina. Although the city itself did not receive the degree of criticism that municipalities closer to the eye of the storm received, the prevailing public opinion that the government had failed the local citizens motivated the state of Texas in general and San Antonio in particular to pay greater attention to emergency management and to ensure that, in the event of another Hurricane Katrina–like disaster, the city was well prepared and equipped to respond. As a result, when Hurricane Dean threatened the Texas coast, San Antonio was fully prepared to receive evacuees and provide other forms of disaster assistance, and, as one interviewee noted, federal assets "were so far forward they almost fell on top of us."

Common Guidance. An additional benefit of grants is that they often include requirements that promote a common operating approach, such as adherence to NIMS principles. Yet, agencies do not need to receive grants to enjoy such benefits. Medical and public health professionals at each location reported relying on guidance and directives from CDC, and both military personnel and civilian actors commented on how the NRF helped both sides move toward all-hazards planning. The acceptance of NIMS principles, including plain talk and the Incident Command System (ICS), was also cited as ways for military and civilian organizations to work together more effectively. However, some interviewees noted that, although improvements had been made in this regard, differences still existed between how the military and civilian sides implemented common guidance like the Emergency Support Function (ESF) component of the NRF.

Regular Interactions. Regular interactions among military and civilian stakeholders in the form of meetings and events had favorable implications for local disaster preparedness efforts. At times, grants served as a catalyst, given their schedules of mandatory exercises and other collaboration-oriented requirements. Regular meetings were another consequence of funding awards. In San Antonio, interviewees felt that grant money was probably the impetus for the city's successful practice of weekly emergency management–focused meetings. Events in the form of both exercises and real-world responses were also perceived as promoting more effective collaboration and communications. Even tabletop exercises provided different organizations with an opportunity to better understand one another's capabilities and to identify potential redundancies, conflicts, or sources of disconnect that could hamper a disaster response. Actual events, including major incidents (like Hurricane Katrina) and smaller events with the potential for casualties (like the annual Western Hemisphere Institute for Security Cooperation [formerly the School of the Americas] protest in Columbus, the National Collegiate Athletic Association [NCAA] Final Four basketball games in San Antonio, and annual New Year's Eve festivities on the Las Vegas Strip), provided civilian and military stakeholders with an opportunity to plan, exercise, and even respond in a complementary, integrated fashion.

---

that increases in the team productivity occurred not because of any modifications to the working environment but rather due to the attention the workers received from the researchers.

Social Relationships. Many interview participants viewed putting faces to names; developing informal connections, including connections across civilian-military lines; and networking as critical to effective disaster planning and response. For example, one Norfolk-area interviewee said that 90 percent of emergency preparedness is networking and communicating with the people who have what you need; another interviewee in Columbus claimed that social capital is the "backbone" of disaster preparedness. In a related vein, several of the emergency management personnel with whom we spoke viewed their informal network as one of their organization's most important assets or capabilities. As one emergency management professional in the Las Vegas area put it, "I rely on face-to-face communications, personal relationships, and rapport. When push comes to shove, I am going to go to someone I know and have a relationship with." He regarded his carefully constructed network as an asset that afforded him unique knowledge of less readily apparent capabilities that other agencies had. In a similar vein, a Columbus-based civilian tasked primarily with emergency management and homeland security responsibilities asserted that he was the only person who could handle coordination in the city, in part because of his networking efforts.

Some interviewees maintained that there was no substitute for social relationships that were both professional and personal—ones that had individuals on a first-name basis and comfortable with calling one another at home during off-hours. However, others tended to focus more on building relationships through face-to-face meetings and regular communications so that one is well aware of "who knows what" within the community and knows whom to seek within an otherwise vast, faceless organization.

Legal Agreements. Military and civilian interviewees from all five sites indicated that memoranda of understanding (MOUs), MAAs, and other formal agreements documenting the roles and responsibilities of key stakeholders were key to effective collaboration and coordination, particularly during disaster response. These memoranda and other agreements tended to be function-specific and were in place to guide the interaction not only between civilian agencies but also between military and civilian functional equivalents. For example, in San Antonio, the civilian hospitals, the local VA medical center, and the military hospitals were parties in agreements that characterize each facility's role and obligations in everyday responses and mass casualty events. In Tacoma, the medical memoranda of agreement (MOAs), MAAs and MOUs involve not only the medical facilities but also the local county health departments, and they provide a basis for coordinating SNS exercises and developing a plan of action for pandemic influenza.

Agreements were also in place between law enforcement agencies and fire departments. The codified relationship between fire departments typically called for a relatively high level of integration, with military fire companies responding to calls off the installation and, to a lesser extent, civilian fire assets providing support for incidents and potential incidents on military property. Given the limitations imposed by the Posse Comitatus Act and other legal and policy strictures, the agreements between military law enforcement and civilian law enforcement agencies were somewhat different from agreements with fire departments, often pertaining to specific assets, such as Special Weapons and Tactics (SWAT) teams. While at least one interviewee expressed a preference for more specific coordination details determined in advance over these relatively generic agreements, overall, the individuals who discussed MOUs and other legal arrangements felt they had favorable implications for disaster preparedness and response.

Information and Communications Technology (ICT). As we discuss in the next section, "Barriers," interview participants tended to emphasize how ICTs could hamper disaster planning and response. Yet, across locations, they also offered examples of ways in which ICT facilitated disaster planning and response. Some of these examples, such as Raytheon's ACU-1000 interconnectivity system and mobile communications trailers like the ones maintained in San Antonio, Columbus, and Las Vegas, may be viewed as strategies to avoid ICT interoperability challenges. In San Antonio, WebEOC® was referred to as a "common denominator" that helps provide situational awareness across functions within a single military installation and for civilian agencies. Interview participants in other locations, like Tacoma, spoke favorably of WebEOC as well. Also in Tacoma, although concerns were voiced about the military's ability and willingness to use a civilian agency-provided tool, the Pierce County Department of Emergency Management portal, civilian interviewees tended to laud it as a database containing information about upcoming training events, classes, and exercises; a repository for after-action reports (AARs); an inventory of every civilian resource available in Pierce County; and a source of additional, potentially helpful information, such as school blueprints and emergency exits.

In another location-specific application, Las Vegas–area interviewees noted that the LEPC relied on a wiki, a computer-based collaborative tool, to facilitate the planning process across representatives from different agencies. Specifically, the Clark County Office of Emergency Management and Homeland Security would draft a plan, post it on the wiki, advise LEPC members that the plan was available for review and comment during a specified time frame, and, after comments from LEPC members were posted and considered, a revised version of the plan was adopted. One LEPC member we interviewed regarded the wiki system as a fantastic tool that saved members from having to attend numerous meetings as a plan was being developed. A more comprehensive list of electronic and other decision support tools, including those used at the locations we visited, is provided in Appendix H.

Barriers.

ICT. As the preceding discussion suggests, by far, the most–frequently mentioned obstacles to disaster preparedness—and response—were ICT shortcomings that precluded essential communications among the many entities involved in a potential response. Some of the ICT-related barriers pertained to a lack of a common operating platform; all the communities we visited noted interoperability problems related to different equipment and computer software. For example, different organizations, including first responders, often possessed communications equipment that operated on different radio frequencies or were otherwise incompatible. Sometimes, incompatibility existed because agencies had selected competing products (e.g., radios by Motorola and Maycom both used in the Las Vegas area); at times, it stemmed from use of different generations of equipment. In one such instance, the city of Columbus had upgraded its equipment, but equipment owned by neighboring cities in the same county was substandard in some cases.

Software products also varied among organizations, and these differences impeded the exchange of important information among them, as did access limitations, such as computer firewalls. For instance, also in Columbus, the county Public Health Department and Fort Benning's military hospital used WebEOC, but the local civilian hospitals relied on LiveProcess, a competing brand of emergency management software. Further, even when the same type of emergency management software was used, typically WebEOC, different versions of it were used by different organizations, and customized local versions of the software were not linked with one another. Accordingly, some interviewees viewed software like WebEOC not as a facilitator but as a barrier because not all stakeholders were using it. With respect to firewall-related limitations, these often stemmed from protections in place for military computer networks. By way of illustration, Fort Lewis personnel noted that the Pierce County emergency management portal, a Tacoma-area emergency management database, could not always be accessed, in part due to Army ICT firewalls and other restrictions.

Other software issues raised included those related to the electronic medical record systems. For example, DoD's software for military health records, Armed Forces Health Longitudinal Technology Application (AHLTA), and the VA's system are not yet fully interoperable.

These ICT impediments posed problems for communications and coordination among different civilian or military functional organizations as well as between local military installations and civilian agencies more broadly. As noted earlier, civilian-civilian communications in Columbus were hampered because the county Public Health Department and local civilian hospitals used different software; in a related vein, Norfolk-area Navy personnel explained that military-military communications were hindered because different Navy computer systems and different email domains were not always connected. With respect to military-civilian interoperability, Nellis AFB personnel noted that, while the installation had some radios that could communicate with civilian equipment, they were unable to obtain licensing authorization to use civilian radio frequencies. On the flip side, one San Antonio interview participant discussed how DoD radio channels were all restricted.

Interview participants discussed a number of strategies and work-arounds they had developed to circumvent these obstacles. Some were relatively crude or simple: In Columbus, interview participants said they relied on telephone calls between individuals to determine bed counts and available blood units because WebEOC and LiveProcess did not "speak" with one another; in Tacoma, interview participants discussed using ham radios to overcome radio-frequency differences; and, in more than one community, civilian organizations purchased equipment for use by military personnel to ensure that all responders could communicate. More sophisticated solutions included the use of universal translators, such as the ACU-1000 or mobile communications vehicles.

Drills and exercises expressly focused on identifying communications deficiencies were also cited as helpful in this regard. For example, in Las Vegas, different agencies came together during "communications rodeos" to develop solutions to IT challenges and other communications issues that emerged in various scenarios. Overall, interview participants felt that progress had been made in overcoming ICT-related obstacles to communications, but barriers still remained.

Lack of Common Terminology. Effective communications among the agencies and actors tasked with disaster planning and response were further stymied on occasion by a lack of common language. Individuals from multiple sites noted that there were codes and acronyms that did not translate across functions or across military-civilian boundaries. As one civilian interviewee put it, the military "is drowning in acronyms," and law enforcement agencies in particular tend to assign different meanings to the same code; a "10-13," for instance, may not denote the same type of incident across jurisdictions. Likewise, a representative from a civilian fire department noted that, while fire officials interpreted the phrase "charge the line" to mean "put water in the fire hose," to a police officer, it instead means "fire away." This was potentially an even greater problem between military personnel and civilian actors. In Norfolk, an interviewee noted that civilians use color-coding for threat levels while the military uses force protection conditions alpha, bravo, charlie, and delta to denote the same. Military and civilian medical personnel also used different terminology at times, which resulted in triage-related problems during at least one exercise: Military personnel employed battlefield triage tags at the incident site, but, when the civilian medical center received the mock patients, its personnel had to re-triage the patients because they could not understand the battlefield tags.

Strategies used to overcome this barrier include the increased reliance on "plain talk" (i.e., forgoing the use of codes for common English phrases) across functions and by agencies. As discussed in the "Facilitators" section, the movement to structures and procedures that are compliant with NIMS, which includes the use of plain talk in radio communications, also tended to help in this regard. Lastly, employing individuals with military experience in civilian agencies was viewed as extremely useful; such on-site "interpreters" not only help with acronyms but also provide an understanding of military command structure and culture.

Information Safeguarding Practices. During interviews, we learned that classified information, which is sensitive in nature, might not be available to individuals who need the information to inform their plans, risk assessments, and exercises. This was particularly a concern for intelligence sharing, and pertained not only to information classified as secret or higher by DoD but also to civilian agency-produced information. Threat assessments conducted by military personnel were typically classified as secret or even top secret, for instance. On the civilian side, interview participants noted that law enforcement agencies did not always share their critical infrastructure assessments with other civilian organizations, and some interviewees thought that public health personnel were reluctant to share epidemic-related data.

To combat these issues, some civilians with emergency management responsibilities obtained a DoD top-secret clearance, and, on occasion, installation ATOs developed a less-sensitive version of their threat assessment briefing that could be shared with civilian law enforcement agencies and others involved in intelligence gathering and counterterrorism efforts. In Las Vegas, the DHS-funded Fusion Center expressly focuses on unclassified local information, in part because it complements top-down classified information to which Washington, D.C.–based analysts were privy and in part because representatives from all agencies—local, state, and federal—can readily aggregate and analyze information of this nature regardless of their security clearances or lack thereof.

Personnel Turnover. We often heard that lack of continuity among the personnel tasked with disaster preparedness or emergency management responsibilities posed challenges. This issue pertained in particular to local military personnel; due to deployments and permanent change-of-station moves, it was difficult for installations to sustain an emergency management–related knowledge base and for employees of civilian agencies to know who is in charge of particular functions on the installation. Personnel turnover in the emergency management function made collaboration like exercise and long-term programs more difficult, and even turnover at the installation commander level every two or three years was cited as a barrier to disaster-related coordination. One interviewee also indicated that turnover of hospital employees could be problematic and that it was difficult to retain individuals trained to use all the communications equipment.

The main strategy discussed to counter this impediment on the military side was the use of civilians to fill emergency management positions rather than active duty military personnel. At every installation we visited, we observed DoD civilian employees filling positions in functions related to plans, exercises, civil engineering, antiterrorism, medical, and fire, among others. For example, the installation-level emergency managers at Navy locations in the Norfolk/Virginia

Beach area, civil-engineering personnel at McChord AFB, and several of the ATOs we interviewed were all civilian employees.

Legal Constraints. At each location, individuals mentioned legal challenges to disaster planning and response. The most commonly cited legal constraints were military limitations imposed by—or perceived to be imposed by—federal legislation (such as the Stafford Act and Posse Comitatus Act) and DoD rules, primarily in the DSCA policy. These mandates restrict the military's ability to provide support for local disasters: the Stafford Act and DSCA limit military personnel to actions in the local community that save lives or prevent immediate property damage; the Posse Comitatus Act may provide criminal penalties for using military personnel for law enforcement activities off the installation, unless those activities are otherwise authorized by statute. Both military and civilian interviewees discussed how these legal constraints reduced the potential role of the local military installation in a disaster response. On a related note, we heard during interviews that the procedures for requesting or approving support for DSCA were unclear.

Lastly, military personnel at McChord AFB noted a different type of legal impediment: regulations that limit the extent to which DoD civilian employees can be assigned active duty military tasks. Specifically, a civilian employee can perform only those tasks that are part of his or her formal job description. At McChord AFB, this meant that DoD civilian employees, even those working in medical units, could not do decontamination and could not serve on medical disaster response teams.

Interviewees did not offer many strategies to overcome these hindrances. We learned that military personnel were not prohibited from supporting disaster relief efforts on their own time, so local military installations could—and did—informally provide personnel-based support in the wake of a disaster before federal support arrived or DSCA-related approvals were received. In addition, in Las Vegas, a reference was made to pending legislation that would enable local civilian agencies to access local military and National Guard assets more readily, but we were unable to verify this independently.

Resource Availability. Lack of resources was a common theme across locations, although the type of resource varied. Personnel-related problems were cited again, but, in this context, the issue was not turnover but rather a lack of time or manpower that could be dedicated to disaster planning and response-related responsibilities. On the military side, emergency management responsibilities were frequently collateral duty, and "multi-hatted" individuals found it challenging to devote the time necessary to perform all of their various duties. Staffing constraints were felt in civilian agencies as well; one of the communities was being encouraged by DHS to establish a Fusion Center, but law enforcement officials were concerned about doing so without compromising more basic law enforcement responsibilities.

Inadequate time was also discussed in the context of scheduling constraints. Although many agencies saw the value in multifunctional and even military-civilian exercises, they were stretched too thin meeting their own exercise requirements and attending to other responsibilities to have the time to devote to such collaborative efforts. Further, even when large-scale exercises were planned or invitations accepted to participate in another organization's exercise, sometimes response to real events precluded participation.

Lack of funding was also cited as a resource-related barrier. Funding was sometimes unavailable for equipment that would improve interoperability, for instance, and budgets did not always permit agencies to send their staff to training or contribute resources to exercises (in terms of people and equipment) as they would have liked. External funding, such as grants, helped civilian agencies somewhat in this regard, but military installations did not receive this type of support.

Perceptions. A barrier of a very different nature, yet one that merits reporting given its potential influence, pertains to individuals' perceptions. Interviewees occasionally expressed concerns that other key stakeholders in local disaster preparedness did not accurately perceive the context in which disaster planning, exercises, and response occurred. For example, we heard civilian interviewees state that they could not count on much support from their neighboring military installation and that the installation would be "locked down" if terrorism was involved, but personnel at one installation asserted that civilians were incorrect in this view. Rather, they indicated that the military would often be able to make assets of some sort available for most disasters.

Conversely, military personnel at other locations noted that, in the past, civilian agencies and even private citizens assumed that the local military would provide extensive, multifaceted support following an incident, and efforts were undertaken to educate the community about what the military could and could not do. Civilian interviewees were also aware of how barriers of a more psychological nature may impede collaboration. In one such case, a civilian interviewee noted that increased security on installations made it more difficult for civilians to get on the installation, which contributed to a perception that the military was more isolated from the community than it was in reality.

Summary

In Chapter Six, we describe the framework for the planning support tool, based on the policy context described in Chapters Two and Three and what we learned from our interviews from the five sites, as described in this chapter. Although our data-collection efforts were bounded by both project scope and the availability of personnel at our selected sites, analysis of site visit data yielded important insights that provide a firm foundation for that framework—one that is supported both by a careful review of current disaster preparedness guidelines and a field-based study of local disaster preparedness practices. Moreover, findings gleaned from our field work have sharpened our focus and have underscored the importance of creating a tool that will both enhance the facilitators of better cooperation and collaboration and attempt to overcome as many of the identified barriers as may be feasible. The following key findings emerged from our analysis of the site visit results across the five sites.

Definition of Community

Across all five sites, civilian and military leaders gave fairly consistent definitions of what the "community" comprised and what constituted their own boundaries for disaster planning purposes. Civilian interviewees generally had a more expansive view of community that included the main city and county and, in some cases, neighboring counties or districts. Their definitions tended to be bounded by where the population lived and worked. Civilians also viewed the military installations as largely independent from the city, primarily due to limitations on when military personnel could cross the boundaries of the installation and limitations on when civilians could enter the base, and perceptions especially on the part of civilian planners of legal restrictions about aid that the installations could provide to the surrounding community. Military leaders tended to define the community in terms of what they were responsible for in an emergency, which was mostly inside installation boundaries (i.e., "inside the wire"). Nonetheless, we did hear about various ways in which agencies interact with one another, including interactions across military-civilian lines. This is discussed in the next chapter, and it forms the basis for one element of the planning support tool described in Chapter Six.

Planning and Exercises

Military installations and local civilian planners both approach major disaster planning from an all-hazards perspective. Nevertheless, both the process and the end product may vary by installation and military service and by community. Planning usually starts with a risk assessment, including threat and vulnerability assessment. Exercises (tabletops, functional drills, and larger-scale field exercises) are used to test and refine plans and to meet external requirements. In principle, all relevant local agencies should coordinate their planning and exercises, but findings from the five sites we visited suggest that there is room to improve local coordination.

Risk Assessments and Capabilities-Based Planning

Risk assessment for a military installation is a broadly standardized and annual process because all installations are required to meet established DoD benchmarks for antiterrorism protection, among other hazards. However, the process varies substantially across sites, depending on the key players. Most military risk assessments are not shared with the civilian community. DoD provides a number of tools to help users conduct risk assessments. The most commonly reported of these is the JSIVA program.

The civilian community has a similar conception of risk assessment, but the process is looser and less standardized. Civilian agencies have also developed a number of tools to address their own needs for a risk assessment template; the Hazard Vulnerability Analysis tool developed by Kaiser Permanente for medical facilities is widely used by civilian planners and by medical personnel on military installations. Without explicitly indicating so, the Kaiser Hazard Vulnerability Analysis assessment tool incorporates not only traditional components of risk assessment but also broad assessment of current capabilities (preparedness, internal and external response).

Facilitators and Barriers to Local Disaster Preparedness

Interviewees reported that facilitators to local disaster preparedness planning included receipt of external funding, primarily federal government grants passed down through the states; attention from external stakeholders, including federal authorities, the media, and the general public; common guidance, such as directives derived from the NRF and CDC; regular interactions among military and civilian stakeholders in the form of meetings and routine or unexpected events; putting faces to names, developing informal connections, and networking; MAAs, MOUs, and other formal documentation of the roles and responsibilities of key stakeholders; and information and communications technologies.

Barriers to disaster planning include shortcomings in information and communications technologies, such as lack of a common operating platform and interoperability problems that preclude essential communications among the many entities involved in a potential or real disaster response, especially across military-civilian lines; lack of common terminology (again, especially terminology used by military and civilian agencies); practices for safeguarding information; lack of continuity among the personnel responsible for disaster preparedness or emergency management, especially on the military side; perceived legal constraints that reduce the military's ability to provide support for local disasters; lack of resources; and inaccurate perceptions on the part of both civilians and the military about what each could contribute in the case of a disaster.

CHAPTER FIVE

Local Emergency Preparedness Networks

Federal guidance and information we gained from interviewees at the five sites highlight the importance of multiagency engagement in disaster preparedness. Whether through formal requirement or spontaneous efforts to meet perceived local needs, or both, some local agencies are working together to coordinate their disaster preparedness planning and response efforts. Yet, interviewees in nearly all the communities represented in our site visits saw opportunities for better local networking.

One of our objectives related to local networking was to begin establishing a baseline against which to measure the future growth of local inter-organizational partnerships—with a particular emphasis on civilian-military cooperation. We imagined that certain organizations were more likely to have established strong ties already with multiple partners. We were interested in seeing which were the most active or influential organizations in local networks from among the main functional areas of responsibility described in the "Data Collection" section of Chapter Four and summarized in Table 4.2 in that chapter: emergency management/planning; public health or medical; security or law enforcement (including antiterrorism); or fire and EMS (including CBRNE and HAZMAT).

In addition, we hoped to learn how we might support further evolution of local preparedness networking by developing a social network component for the planning tool. Therefore, a second objective related to local networking was to use social network analysis (SNA) to help elucidate the structural factors that either facilitate or hinder local cooperative preparedness.

We begin our discussion with a brief background on the science of SNA and its potential application to local networking to improve disaster preparedness, then describe the survey we conducted and highlight our findings across the five sites. As described later in the chapter, the network data we present are provided only as initial proof of concept because incomplete data availability limited the robustness of our analyses. Findings from each site are described in more detail in Appendix G. We conclude the chapter with a summary of our findings and suggested implications for the RAND planning support tool described in the next chapter.

Background on Social Network Analysis

Theory and Traditional Uses

The basic assumption of SNA is that the structure of relationships among a set of actors, including the location of specific actors within a network, have "important behavioral, perceptual, and attitudinal consequences both for the individual units and for the system as a whole" (Knoke and Kuklinski 1982, p. 13; see also Wasserman and Faust, 1994). SNA helps inform observations about the benefits and disadvantages of various network structures and thus suggests constraining and enabling dimensions of relationships among actors.

Network analysis can focus on the positions of each network node (which may represent an individual, a group of individuals, or an organization) on the network or group as a whole, or both. Thus, SNA approaches underscore the fact that the nature of the network or group is determined by the structure of relationships between the nodes, while the behavior of an individual node is partly defined by the nature and structure of its group affiliations (Emirbayer and Goodwin, 1994).

In a review article, Ressler (2006, p. 2) argues, "Network analysis seems to work because it provides a structural analysis while still leaving room for individual effort." Although SNA does acknowledge that an individual node may exert its own unique influence, the structure of the network (more or less centralized or more or less hierarchical, for example) and the nature of an individual node's links to other nodes in the network are more relevant for our purposes within the context of this study and thus are the focus of our analyses: The sum of the individual relationships between member nodes determines the structure and behavior of the larger network.

Network analysis has long been used to explore the structural determinants of social phenomena, with an emphasis on one of the central functions of networks—access to information (Granovetter, 1973; Wellman and Berkowitz, 1988; Monge and Contractor, 2003). Granovetter (1973), in one of the most influential pieces of SNA research, found, somewhat counter-intuitively, that "weak ties" (i.e., relationships with acquaintances) are more important than "strong ties" (i.e., relationships with family and close friends) when one is trying to find a job, or anything else requiring access to information or resources that one does not already have. Weak ties are more useful because weak ties function as bridges between networks. Acquaintances act as bridges between the members of different networks, thereby enabling or enhancing diffusion of information and resources.

In contrast, for a typical small, densely interconnected network, one's close associates might be more *willing* to help provide information or identify new opportunities, but they are actually less *able* to help because they cannot provide any *new* information, resources, or opportunities. Members of a dense network spend their entire relationship budget[1] on one another and do not, therefore, have many ties to individuals outside of the network. Everyone in a dense network shares the same information. This might be positive in the case of an emergency response network operating together efficiently from the same information, but having access to the same information is positive only as long as no new information is needed to solve a never-before-seen problem. More dispersed, non-redundant networks could supply the information seeker with access to a greater diversity of knowledge and information from external sources about a wider array of opportunities. Also, in general, weak ties may be less costly to maintain, since relationships with acquaintances tend to require less time and energy. Individuals with more weak ties may also be able to engage in a larger number of networks, since they spend less of their relationship budget maintaining the ties in any one network, though there may be costs associated with individuals adjusting to each different network.

A second influential element relevant to our work is Ronald Burt's (1992) concept of structural holes. Burt notes that an individual can act as a bridge between two otherwise uncon-

[1] Varda et al. (2008) use this term to refer to the amount of time any one individual has available to both maintain his or her existing relationships and seek new ones.

nected individuals—spanning the *structural hole* between them. This boundary-spanning individual plays an important role in controlling the flow of information between the two individuals; when those individuals are in different networks, this "broker" can facilitate information sharing from one network to another, thereby creating the potential for innovation. A police officer who is also a member of the Guard, for example, will be able to share details about military tactics, techniques, and procedures with his law enforcement network and vice versa. Tactics he learns in one network may be considered innovative when he shares them with the second.[2]

Wasserman and Faust (1994) explain how SNA can be used to distinguish the varying roles of individuals in groups—simply by examining the structure of the group—and to identify and compare subgroups within the larger network. In a more recent popular publication, Malcolm Gladwell (2000) used basic network terminology to describe three types of people critical for understanding the way social epidemics, or fads, begin and spread. Gladwell calls the three types connectors, mavens, and salesmen (see also Milgram, 1967). Both an individual's position within the network and the nature and number of his or her ties to other network members have implications for that individual's behavior and for that individual's ability to affect the behavior of others. Network analysis can help identify individuals in positions of influence and leadership within a network by measuring the characteristics of a node. Fernandez and McAdam (1988), for example, found that individuals who were more central to a given network exerted a stronger influence on the network. However, the more central node also experienced a stronger influence from the network.[3]

Application of Network Analysis to Public Health

In public health, network approaches have also been important both in stopping the spread of infectious diseases and in providing better health care at the community level (Levy and Pescosolido, 2002).[4] Three recent health-related studies are particularly relevant to our efforts to develop a local disaster planning support tool. The first study uses SNA to diagnose a perceived problem: why life-saving vaccines are so slow to be implemented in developing countries (Conway, Rizzuto and Weiss, 2008). The second uses SNA to understand how to improve the effectiveness of local public health emergency preparedness networks in Missouri (Harris and Clements, 2007). The third is an effort to map federal inter-organizational coordination in emergency response networks and compare ideal to actual networks (Kapucu, 2005).

Conway, Rizzuto, and Weiss (2008) used SNA to explore how vaccination programs are established in developing countries. The authors hoped to understand why it takes so long to implement a program that would bring life-saving vaccines to a vulnerable population. Full implementation of such a program has taken as long as 20 years (p. 1). They mapped the national and international vaccine-related networks in four cases, focusing on the relationships and roles that connect the myriad national and international stakeholders (from the World Health Organization [WHO] to nongovernmental organizations [NGOs] to the national government) to each other and to the vaccination projects.

---

[2] Innovation does not require a brand new idea; it requires only that an idea be new to a specific setting.

[3] *Prominence* is a measure of central tendency, evaluating the degree to which some individuals within a group are linked to a larger number of people than the average member, and have more important relationships with those people (Fernandez and McAdam, 1988, p. 365).

[4] For a review of the health research that has employed SNA, see Smith and Christakis (2008).

One of the problems they identified was the failure of those desiring vaccination programs to include their country's banking and finance representatives early in the process. Moreover, they found that, while international organizations, such as WHO and the United Nations Children's Fund (UNICEF), played an important role in the vaccine introduction process, these organizations failed to share information adequately about the vaccine introduction process across countries (Conway, Rizzuto, and Weiss, 2008, p. 3). Ultimately, they concluded that showing stakeholders the network maps can "help stakeholders understand how to improve decision-making processes and thus hasten" implementation of vaccine programs (p. 1). This conclusion is consistent with how the RAND team hoped that SNA of local civilian and military emergency response networks could potentially supplement the site visit interviews and eventually be developed to complement local planning. We feel that, by seeing their own networks, emergency planning stakeholders may be able to identify missed opportunities for connections or identify redundant connections that could be eliminated to increase their relationship efficiency.

Harris and Clements (2007) use SNA to examine relationships among public health emergency planners in Missouri. They looked at communications flows across the emergency management community and found that most planners communicated regularly with their counterparts in the local region, but rarely with public health planners located outside their region. Planners listed an average of 12 local organizations (e.g., emergency management, hospitals/clinics) in their emergency preparedness networks but rarely included federal-level or private-sector contacts (p. 488). Harris and Clements concluded that the emergency planning network could be strengthened by addressing two gaps: (1) increasing the degree to which local planners regularly communicate with non-planners in the network, and (2) establishing increased connections to private-sector representatives (p. 494). Such considerations were relevant to the SNAs we undertook as part of this study, as we discuss later in this chapter.

The third study (Kapucu, 2005) compared the formal emergency response network established by the Federal Response Plan (FRP), subsequently superseded by the National Response Plan and NRF, to the actual interactions of emergency response participants—as reported in post-9/11 FEMA situation reports and supplemented by interviews with participant organizations (p. 34). Kapucu found that actual interactions were much more limited than the densely connected network envisioned in the FRP, tending to occur primarily between organizations of similar types (grouped into public, private, and nonprofit) and remaining within a single jurisdiction (p. 40).

Although Kapucu's effort was primarily descriptive rather than prescriptive, he identified two opportunities for achieving greater efficiency and productivity in actual emergency response networks: increased collaboration among organizations and increased collaboration across jurisdictions. Kapucu's recommendations are based on the assumption that more collaboration across organizational types and jurisdictions will increase efficiency of communications and coordination during an emergency. His assumption is in line with findings from organizational theory that, as organizations increase in size and number of components, they move toward a more bureaucratic structure and adopt SOPs to increase efficiency and coordination (Donaldson, 1995; Child, 1984, 2005).

The structuralist approach embodied by network analysis establishes several key insights. First, network structure can elucidate important details about the positions of the individuals who comprise it. Network structure can also provide insight into how the network operates, indicating whether it is hierarchical or decentralized. Second, comparing the overall structure of a given network to both similar and contrasting networks can help identify anomalous networks (Banks and Carley, 1994), identify the central tendency within a set of networks (Carley, Lee, and Krackhardt, 2001) and characterize desirable networks in measurable terms. Because we know the implications of different types of network structures in terms of communications and planning efficiency, behavior, and susceptibility to breakdown (discussed in further detail in the "Survey Methods" section), we can predict certain network behaviors based on examination of a network's structure (Merrari, 1999; Arquilla and Ronfeldt, 2001; Carley, Reminga, and Kamneva, 2003; Ritzer, 2004; Enders and Sandler, 2006; Enders and Su, 2007) and recommend alternative structures based on the network's specified goals.

We now discuss the goals of emergency management networks and the measures used to characterize these networks and the achievement of their goals. As noted previously, our main focus is on the characteristics of the network as a whole, more than the characteristics of the individual nodes.

Emergency Management Networks and Preparedness

The goal of emergency management is to protect the community "by coordinating and integrating all activities necessary to build, sustain, and improve the capability to mitigate against, prepare for, respond to, and recover from threatened or actual natural disasters, acts of terrorism, or other man-made disasters" (FEMA, 2007b). Certain network qualities render this goal easier to achieve in all phases of the disaster cycle. These qualities include, at a minimum, efficiency, flexibility, resiliency, and redundancy. The categories are not discrete; one category can and does influence the next.

- An *efficient network* is one that communicates new information in the shortest amount of time to the largest number of nodes. It is also one that maximizes the relationship budget of each member node so that members spend an appropriate amount of time maintaining relationships and an appropriate amount of time attending to their other responsibilities (Varda et al., 2008). Finally, an efficient network makes it easy for members to coordinate with one another.
- A *flexible network* is one that responds rapidly and effectively to new information and new challenges. This may occur because of efficient communications, or because the organization is set up to be scalable, increasing in size—by adding reservists, for example, as the number of casualties rises. It may also occur because the structural holes in the network provide a space in which new relationships can be established or through which new ideas will emerge from members at the periphery of the network.
- A *resilient network* is one that can lose a few key members without being completely destroyed. A resilient network is also one without single points of failure. This sometimes means that people are substitutable: They may have been cross-trained to take on each other's responsibilities. For example, rather than having the local emergency manager be the single point of contact for all of the local military installation emergency managers, emergency managers should also have direct ties to other key network players, such as the local police department and public health organization. This ensures that, if the local emergency management organization is attacked, loses power, or loses communications, installation emergency managers will still know what is going on in the civilian community through their redundant ties. Redundant ties are another way to improve the resil-

iency of a network. Flexibility can also contribute to resiliency. A scalable organization is often a more resilient one.

- *Redundancy* is a state in which duplication of a relationship, a system, or a responsibility is established to prevent the failure of an entire system upon the loss of a single component. Redundancy in the context of resiliency is a positive network characteristic, but redundancy can also be a negative characteristic if too much redundancy offsets efficiency. This might be the case, for example, if the local emergency manager not only had ties to all of the installation managers and to other key players on the installation, such as the head of security forces and the chief medical officer, but also spent time maintaining ties to sub-organizations, such as the personnel or facility offices at the military medical centers.

We hypothesize that SNA can be used to improve planning for emergency preparedness by enabling assessment of existing networks, and by identifying (and measuring quantitatively and qualitatively) key changes that could improve a network—assuming that we have defined the characteristics of a "desired" or "ideal" network.

Social Network Measures

SNA tools allow evaluation of the quality and strength of the network as well as the network's ability to meet the specific demands of the emergency preparedness mission, based on the previously identified attributes: efficiency (coordination potential), flexibility (innovation potential), resiliency (to disruption), and redundancy. Standard network measures are used to get at these details (Everett and Borgatti, 2005; Scott, 2000; Hanneman and Riddle, 2005). Table 5.1 defines the key network measures we used in our analysis and links them to the network goals of efficiency, flexibility, resiliency, and redundancy. While there are both nodal measures and network-wide measures, for purposes of our study, we focus mainly on the latter.

Node Statistics. SNA often begins by identifying key players in the network—those in positions of influence or leadership. Both an individual's position within the network and the nature and number of his or her ties to other network members (referred to as *alters*), have implications for that individual's behavior and ability to affect the behavior of others within the network. Measuring the centrality of individual nodes helps to identify which nodes are

Table 5.1
Summary of Network Goals and Measures

Goal	Network Measure	Definition	Interpretation
Efficiency (coordination potential)	Normed closeness centralization	The sum of the distances from each actor to all other actors in the network (scale of 0 to 1)	High normed closeness centralization (e.g., >75%), or high density (e.g., >75%), and short distance (e.g., 2–3), and small diameter (e.g., 2–3) may mean that the network can respond more quickly to stimuli and communicate and coordinate more efficiently.
	Density	Ratio of existing connections to all possible connections (scale of 0 to 1)	
	Average path length	Average number of steps between all possible node pairs	
	Diameter	Maximum number of steps between any potential node pair in the network	

Table 5.1—Continued

Goal	Network Measure	Definition	Interpretation
Flexibility (innovation potential)	Density	Ratio of existing connections to all possible connections (scale of 0 to 1)	Lower density (e.g., <25%) may reflect more flexibility to receive input from outside the network—leading to innovation.
	Betweenness centrality	Based on the number of geodesic paths between all possible pairs of actors in the network, betweenness measures the number of times each node falls on one of these paths (scale of 0 to 1).	More nodes with higher betweenness centrality (e.g., >75%) playing a "broker" role in the network may allow for more innovation and hence flexibility by bridging structural holes.
Resiliency (of network to failure of single node)	Normed closeness centralization	The sum of the distances from each actor to all other actors in the network (scale of 0 to 1)	High normed closeness centralization (e.g., >75%) may reflect less resiliency, since the more dominant a few nodes become, the more vulnerable the network is to the incapacitation of those nodes.
	Density	Ratio of existing connections to all possible connections (scale of 0 to 1)	High density (e.g., >75%) may result in greater resiliency, since density implies redundant relationships.
	Betweenness centrality	Based on the number of geodesic paths between all possible pairs of actors in the network, betweenness measures the number of times each node falls on one of these paths (scale of 0 to 1).	A node with high betweenness centrality (e.g., >75%) has significant influence on the flow of information within the network and plays a "broker" role in the network. These nodes are very important to the network but also function as single points of failure, since they have so much influence. Fewer "brokers," or brokers with lower scores, may provide more resiliency.
Redundancy[a] (to enhance resiliency)	Normed closeness centralization	The sum of the distances from each actor to all other actors in the network (scale of 0 to 1)	High closeness centrality (e.g., >75%) means less redundancy, since communications in more centralized networks go to and from the center but rarely around the sides.
	Density	Ratio of existing connections to all possible connections (scale of 0 to 1)	High density (e.g., >75%) implies redundant relationships.

[a] We are currently missing too much information to be able to use a direct measure of redundancy—but we include the measure as an important one for future expansion of an SNA tool set.

the most critical to the functioning of the network and which nodes might be single points of failure for the networks—for example, a node that connects multiple subgroups to one another but whose incapacitation would turn these subgroups or individuals into isolates.

There are several different ways to measure individual centrality, including the two we use: nodal measures of degree and betweenness centrality. *Degree centrality* measures how many direct connections a node has; it indicates the number of other nodes with which each node works or communicates on a regular basis. It is sometimes used as a measure of a node's popularity. The higher the measure of degree centrality for the individual node, the more important that node is likely to be in the wider network.

Betweenness centrality identifies all of the geodesic (shortest) paths between all pairs of nodes in the network and counts the number of times each actor falls on each of these pathways. The degree to which a particular node is the only link between pairs of otherwise unconnected actors establishes that node's betweenness centrality score. A node with a high betweenness score has significant influence on the flow of information within the network and plays a *broker* role in the network.

Broker nodes are very important to the network but also function as single points of failure. An early analysis of communications networks (Hagen, Killinger, and Streeter, 1997) identified measures of closeness centralization and betweenness centrality as "useful for indicating levels of network coordination" (p. 13). We use the nodal measure of betweenness centrality, discussed here, and the network measure of closeness centralization, discussed in the next section.

Network Statistics. Network centralization measures generally indicate the extent to which a network is dominated by just a few nodes or one central node (Scott, 2000). *Closeness centralization* (normed) is a way to measure access (Hagen, Killinger, and Streeter, 1997).[5] It is a measure that can suggest how long it will take for information to spread throughout the network. The higher the measure of normed closeness centralization (on a scale of 0 to 1), the more quickly information will likely spread throughout the network. A highly centralized network may be *efficient*, but it is probably less *resilient* than a decentralized network that might have more redundancy.[6]

Network density indicates the overall level of network integration among organizations and provides a measure of network cohesion (Scott, 2000). *Density* is defined as the proportion of ties that exist in the network relative to the maximum possible number of ties, thus ranging in values from 0 to 1. Measuring network density facilitates understanding how *efficiently* information flows through the un-centralized portions of the network and how likely *innovation* is to arise within the broader network. The higher the density score, the higher the overall level of network integration and cohesion and the smoother the flow of information, but the less likely innovation is to arise within the network.

Average path length measures the average number of steps between all possible pairs of nodes in the network; the smaller this number (measured in whole numbers with 1 being the

---

[5] Closeness centrality is measured as the sum of the distances from each actor to all other actors in the network.

[6] Ideally, the number of cliques and subgroups and the degree of triadic closure within the network should also be measured to assess how much redundancy exists in the network and to evaluate internal communications flows—identifying potential blockages or potential network efficiency in executing missions that require cooperation from the entire network—such as disaster response. When a large amount of data is missing, these measures are less reliable.

smallest), the more *quickly* new information or orders can be passed through the network.[7] *Diameter* indicates the longest distance between any two members of the network, providing an outer bound for how *slowly* new information may be passed throughout the network. Networks with more connections (greater density) and shorter distances between actors should be able to respond more rapidly and efficiently to stimuli. Networks with fewer or weaker connections, and with longer distances between actors, will be slower to respond to new inputs (Hanneman and Riddle, 2005; Wasserman and Faust, 1994; Corbacioglu and Kapucu, 2006). In addition to these quantitative network measures, scientists also used pictorial representations of networks to highlight more graphically their nature and structure. In this study, we use both network statistics and visualizations to portray the local networks we examined.

Table 5.1 summarizes the measures we use to evaluate how well each network is meeting the key emergency preparedness network goals of efficiency, flexibility, resiliency, and redundancy. There is a high correlation among these measures and across the categories. Each of the measures is affected by the others, often inversely; and each of the categories is also affected by the other categories.

Survey Methods

The network data for the formative research described in this report were collected primarily via a paper-based survey (documented in Appendix F) distributed to interviewees at each of the five sites selected for visits (see Chapter Four for a discussion of the site selection process and key findings from the interviews). Interpretation of the results draws on the information we collected during our interviews at those sites as well as on a general interpretation of the network statistics based on findings from the SNA literature.

In each of the five communities, we asked representatives from key emergency management–related functional areas to complete a survey about the composition of their larger network. We hypothesized that certain functions were likely to be more important to a local emergency preparedness network than others. Therefore, though all of those we interviewed were asked to complete the survey, the team made repeated and focused efforts to obtain returned surveys from target interviewees representing all five mutually exclusive categories of functional responsibility that we created for classification purposes for this study: emergency management, public health or medical, security or law enforcement (including anti-terrorism), fire and EMS (including CBRNE and HAZMAT), and planning or higher-level management functions.

Respondents were specifically asked to identify the "most important" members of their emergency response network. For each nominated network member, respondents were then asked a series of additional multiple-choice questions to characterize their relationships to the other network members. We asked multiple-choice questions about (1) the type of hazard for which planning most often occurred (e.g., natural disaster or terrorism preparedness); (2) the type of collaboration (e.g., joint planning, communications, contracted services); and (3) the

---

[7] The number of steps between each possible pair of actors can also be measured without averaging the number of steps. The number of actors at various distances from each actor could provide us with information about the constraints and opportunities different actors have based on their position in the network, and information about distance can also tell us which actors are the most costly to reach (Hanneman and Riddle, 2005, Ch. 7).

frequency of interaction. A breakdown of survey respondents by community, personnel type (military/civilian community/VA), and functional representation is provided in Table 5.2.

Based on connections reported by survey respondents identifying key members of their emergency management and response networks, we used the surveys from each site to construct a list of emergency management–relevant organizations. The team recorded the relational data in a series of Excel spreadsheets, which were then imported into UCINET version 6 to quantify network attributes; NetDraw was used to create network diagrams.

We developed a short, 12-question survey to maximize the number of respondents who would agree to complete it. Respondents were instructed to photocopy additional pages if they wished to provide information on more than 12 key connections. Seventy surveys were completed and returned to us. In those surveys, the average number of connections listed was

Table 5.2
Summary of Survey Respondents, by Site and Functional Area of Responsibility

			Number of Interviews, by Function				
Area	Overall No. Participants	Overall No. Surveys	Emergency Management/ Preparedness	Security/Law Enforcement	Fire/EMS	Health/ Medical	CBRNE/ HAZMAT
San Antonio, Texas, metropolitan area							
Civilian	3	2	1	0	0	1	0
Military	14	13	5	2	1	5	0
VA	1	1	0	0	0	1	0
Norfolk/Virginia Beach, Virginia, metropolitan area							
Civilian	5	5	2	1	1	1	0
Military	8	8	1	3	2	2	0
VA	1	1	0	0	0	1	0
City of Columbus and Muscogee County, Georgia							
Civilian	5	5	2	1	0	2	0
Military	7	7	1	2	1	2	1
VA	1	1	0	0	0	1	0
City of Tacoma and Pierce County, Washington							
Civilian	3	3	1	1	0	1	0
Military	16	12	4	3	1	2	2
VA	1	1	0	0	0	1	0
City of Las Vegas and Clark County, Nevada							
Civilian	5	4	1	1	0	2	0
Military	6	6	1	2	1	1	1
VA	1	1	0	0	0	1	0
Total	77	70	19	16	7	24	4

ten, with a maximum of 34 and a minimum of one. The modal number of connections was 12, with 14 respondents listing more than 12 and 45 listing fewer than 12 connections.

To capture both formal and informal relationships within the local community, we did not give respondents a complete list of possible organizational choices from which to draw their list of partners for disaster preparedness purposes. They were provided an illustrative list of partners across a number of categories, such as federal, state, local, military, and nonprofit (see p. 9 of the survey in Appendix F). Respondents were free to name any organization considered to be a key partner at the local, state, or federal level. In practice, respondents tended to emphasize local and in-state regional organizations, mostly those physically located in their community.

Respondents tended to identify their functional counterparts at lower levels in the organization rather than referring to the larger or parent organization (for example, respondents might identify a Coast Guard unit from a specific district, or they might refer to the Population Health Directorate within NMC Portsmouth). To account for the latter, the RAND team first aggregated lower-level units into larger categories or their parent organizations (e.g., individual schools were aggregated into the larger category of educational institutions), as suggested in the literature (Pentland, 1999). Second, the RAND team assumed that a response linking one partner to another constituted a linkage even if the nominated partner did not likewise name the nominating partner. That is, relationships were made reciprocal for the purposes of this analysis. The team used these data to create a network-analysis matrix for each site.

The survey methodology—in particular, the non–roster-based format in which respondents were encouraged to create their own list of key disaster preparedness partners—was subject to some inherent limitations and biases. In particular, there was likely an underreporting bias in terms of the number of connections listed by each respondent; that is, we expect that our respondents actually work with more agencies than they listed on our survey form.

More importantly, on average, each site-specific network represents some 67 different organizations, but network structure is derived from information collected from just 14 surveys (20 percent of network members). The majority of partner organizations nominated by survey respondents did not have an opportunity to report their own partners. This means that survey respondents are most often the center of their own star-shaped cluster (or neighborhood) within the larger local preparedness network. Star-shaped clusters are those in which a central node is connected to a star-shaped array of other nodes, none of which is connected to another. The nodes attached to the central node but to no other node are called *pendants*. If we were to complete each network by obtaining information about the nominated organizations' ties to other network members, there would be far fewer star-shaped neighborhoods. Thus, the survey methodology had a significant impact on the structure of the four networks we elicited for this proof-of-concept activity.

In addition to these biases identified, the survey methodology led to some key limitations for analyzing the data. One of the most common network measures of influence and flow of information and leadership in a network is the measure of in-degree versus out-degree. *In-degree* is the number of participants that nominate a specific organization or node as a key disaster preparedness partner in their network. *Out-degree* is the number of key disaster preparedness partners that the first node nominated. The bias of our survey instrument makes these measures undesirable. There is no way for us to know, for example, whether node A was excluded from node B's network intentionally, or merely because node B forgot to include node A in the initial nomination process. We did supplement the network analysis with information obtained in the semi-structured interviews at the five sites in order to identify connections between key nodes even if they were not explicitly noted in a respondent's completed network survey—though this was done only very minimally. This included, for example, connecting a civilian emergency manager to FEMA in the same way that the local military emergency manager established that tie.

Another common group of network measures, and ones that we employ despite the aforementioned limitations, are measures of centrality. Survey respondents are almost always going to have the highest centralization scores, certainly higher than non-respondents. We remain fairly confident, however, that the survey respondents we selected represent the core of the network. As such, we believe that the missing data would trend in the same direction in terms of which organizations prove to be most central to each local network. Moreover, in the event that missing data were to reinforce rather than undermine the centrality of the organizations that score highest in our current analysis, then the broader local preparedness network is likely to be more centralized than our analysis currently reflects. However, we are unable to determine whether that is the case, given the design of this initial study.

We are also currently missing too much information to be able to use a direct measure of redundancy—but we include the normed closeness centralization measure in our analysis (using a proxy) as an important measure for future expansion of the totality of network measures. Capture of data that are currently missing will result in a much greater redundancy of ties between network neighborhoods. As a corollary, network density is underestimated and will increase as additional data increase the number of reported network ties.

Because we are fairly certain that the organizations we selected to complete the survey are organizations that actually form the core of each local network, we are confident about our observations related to the core network. We cannot reliably say much about organizations at the periphery of the network, however. Our observations about the network as a whole are also biased by the large amount of missing data—though the observations we do make are in line with our findings from Chapter Four.

The results and analysis presented in the next section are provided as a proof of concept for what can be done with more complete data or with a larger population of respondents. Despite the fact that our analyses are based on biased and very sparse data, they still illustrate the capacity for SNA to enable and improve collaboration. Network visualizations sensitize organizations to the universe of possible partners and help them understand their current location in the network. From here, they can work to improve their collaborative roles. Network indices, taken as a whole, help to identify an organization's location and role in a network. Indices also help to identify the overall capacity of a network to facilitate collaboration and mobilization. Once this information is known, network members can change the structure of the network as a whole or their own location in a network to improve collaborative capacity.

Results and Analysis

This section first reports and describes the statistical measures of local networks across the five sites where the RAND team conducted interviews (see Table 5.3 for a comparison of network statistics across these five sites). We then present a picture and highlight key characteristics of each network (more detailed descriptions of the five networks are presented in Appendix G).

Table 5.3
Comparative Network Statistics Across Sites

Location	San Antonio Metropolitan Area (Texas)	Norfolk and Virginia Beach (Virginia)	Columbus and Muscogee County (Georgia)	Tacoma and Pierce County (Washington)	Las Vegas and Clark County (Nevada)
Network measure					
Size of identified network	68	81	54	86	45
Closeness centralization (%)	33	39	39	41	36
Network density (%)	12	9	11	9	16
Average path length	2.84	2.81	2.77	2.75	2.12
Network diameter	6	5	5	5	4
Node measure					
Highest degree centrality scores	1. San Antonio PH 2. San Antonio EM 3. Randolph SF	1. Virginia Beach EM 2. NMC Portsmouth 3. NS Norfolk EOC	1. Ft. Benning EM 2. Muscogee County PH 3. Columbus EM	1. McChord Plans 2. McChord Medical 3. Pierce County EM	1. Southern Nevada Health District 2. Las Vegas Emergency Preparedness 3. Nellis BEE
Highest betweenness centrality scores	1. San Antonio PH 2. Randolph SF 3. San Antonio EM	1. NMC Portsmouth 2. Virginia Beach EM 3. NS Norfolk EM	1. Ft. Benning EM 2. Columbus EM 3. Muscogee County PH	1. McChord Plans 2. Ft. Lewis Installation Safety 3. Pierce County EM	1. Southern Nevada Health District 2. Las Vegas Emergency Preparedness 3. Nellis BEE

NOTE: PH = public health. EM = emergency management. SF = security force. BEE = bioenvironmental engineering.

Comparison of Networks Across Sites

Most Influential Organizations. Acknowledging potential limitations in our network data based on incomplete responses, our findings aligned closely with our hypotheses about which functions would be most important to local disaster preparedness networks as well as with our findings in Chapter Four. The most influential organizations across all five sites (in both the civilian and the military communities) were consistently emergency management and planning; health and medical; and security and law enforcement offices, as reflected in their high degree centrality and betweenness centrality scores. Organizations with the highest degree centrality scores were the most "popular" nodes in the network. They worked or communicated with the largest number of partner organizations on a regular basis. The higher the measure of degree centrality for the individual node, the more important that node is likely to be in the wider network.

The organizations with the highest betweenness scores are responsible for bridging the gaps between otherwise disconnected nodes. The higher the betweenness score, the more relationships that node is responsible for brokering. Not only are nodes with high betweenness scores very influential in terms of controlling the flow of information in the network, but they are also single points of failure in the network, since their removal from the network leaves pairs of nodes completely disconnected from one another. The most between organization in the San Antonio metropolitan, Norfolk/Virginia Beach, and Las Vegas/Clark County areas was a public health or medical organization. In Columbus/Muscogee County and Tacoma/Pierce County, the most between organization was a military installation emergency management or plans office. The local emergency management office had the second highest betweenness score at four of our five sites. In third place were two public health/medical organizations (one civilian, one military); two emergency management offices (one civilian, one military), and installation security forces at one site. It is important to keep in mind that betweenness centrality is likely overestimated in each of our cases, but we are confident that the scores for our core network members (those who filled out surveys) are fairly accurate.

A few other organizational types came up repeatedly as playing an important role—as reflected in multiple nominations from survey respondents, high degree scores, high closeness scores, or high betweenness scores—including (1) fire services, (2) civil-engineering and public works functions (e.g., local utilities), (3) local businesses and critical infrastructure owners, and (4) educational institutions, such as elementary schools. We discuss these in the sections that follow.

Communications flow, coordination, and innovation. The first measure listed in Table 5.3 is the size of the identified network. We include this number as a point of information and because larger networks tend to have more difficulty than smaller ones in successfully managing communications and coordination across the network (Child, 1984; Marlow and Wilson, 1997; Serenko, Bontis, and Hardie, 2007). This means that it should be easier for the smallest network, Las Vegas and Clark County, to effectively coordinate and manage its communications than it is for the largest network, Tacoma and Pierce County, which is almost twice as large.[8] The size of the identified network reflects the total number of organizations listed by survey respondents at that site as being "most important" to their emergency preparedness networks. The number also includes the survey respondents themselves.[9]

Normed closeness centralization indicates the degree to which each of the networks is dominated by just a few nodes or one central node.[10] It also provides a sense of how long it will take for information to spread throughout the network. A perfectly centralized network (with a score of 1 or 100 percent) is a star-shaped network in which every organization is connected to the center node with no ties between the other organizations. The higher the measure of normed closeness centralization (on a scale of 0 to 1), the more centralized the network is—meaning that the network is controlled by just a few nodes, rather than having a decentralized structure.

---

[8] Keep in mind that, because of the large amounts of missing data and the associated problems mentioned in the "Survey Methods" section, the comparison of network size across sites is not as meaningful as it would be had we been able to standardize survey respondents across sites. Where it is meaningful is in evaluating the site-specific measures, such as centrality and density.

[9] To better contextualize the variation in network size, we again note the number of completed surveys per site as listed in Table 5.2: San Antonio metropolitan area = 16 surveys, Norfolk and Virginia Beach = 14 surveys, Columbus and Muscogee County = 13 surveys, Tacoma and Pierce County = 16 surveys, and Las Vegas and Clark County = 11 surveys.

[10] Closeness centralization is measured as the sum of the distances from each actor to all other actors in the network.

We can also think of normed closeness centralization in terms of access. Higher normed closeness centralization scores imply more centralized access to resources or information. The more centralized the network, the easier it should be for members of the network to get information, since the central manager can transmit the information to each of its partners simultaneously.

The network-centralization scores are similar across all five sites. The most centralized networks are in Tacoma/Pierce County, Norfolk/Virginia Beach, and Columbus/Muscogee County. But their scores (41, 39, and 39 percent, respectively) are not that different from the other two sites, San Antonio metropolitan area (33 percent) and Las Vegas/Clark County (36 percent). These are all moderately low normed closeness centralization scores, indicating that overall centralization across all of our sites is fairly low. Members of a more decentralized network often have difficulty accessing information in a timely manner. This correlates with our site visit finding that planning is most often done separately by civilian and military communities and often stovepiped by function; it appears that collaborative planning has increased in recent years. Moreover, the low closeness centralization scores also correlate with something we heard regularly during interviews about exercises. Interviewees often noted that it was difficult to learn about other exercises being conducted by other organizations in a timely enough manner to be able to participate. Alternately, some interviewees remarked that there were always so many exercises going on that it was difficult to keep track of, or find the time to participate in, all of them. Both of these observations are unsurprising in a decentralized network.

Not all efficient networks are centralized. Densely connected, decentralized networks can be equally efficient. To measure the efficiency of a decentralized network, we measure its density—the degree to which all organizations that can be connected are connected. The higher the density score, the higher the overall level of network integration and cohesion, and the more efficient the transmission of information is likely to be. Network-density scores across each of the five sites are very low, ranging from 9 percent to just 16 percent at the high end. This means that none of these five networks is very densely connected. Although density is likely underestimated in all of our networks due to the large amount of missing data, we believe that the direction of the scores (lower rather than higher) is accurate. This is a disadvantage in terms of communications and coordination efficiency. There are too many holes in the network in which organizations are either isolates (disconnected from everyone) or pendants (connected to only one organization in the larger network and therefore dependent on that organization to pass along all critical pieces of information).

On the positive side, the holes in the network provide opportunities for innovation. Ronald Burt (1992) coined the term *structural hole* to refer to the gap that exists between two different groups (for example, between the military and civilian communities). In his research on creativity, competition, and collaboration, Burt found that individuals who bridge the gaps between social groups are more likely to innovate. Moreover, Burt argues that innovation is merely the successful introduction of a mundane idea to a new setting. Though none of our networks is densely connected, the long-term goal of many of our interviewees was increased collaboration—moving toward more–densely connected networks. As such, these low density scores may emphasize a moment of opportunity in which very loosely connected communities (hereafter, *network neighborhoods*) can borrow the best ideas from one another as they move toward closer collaboration. We acknowledge once again the potential limitations of our findings because we did not have responses from all agencies listed; nonetheless, we are confident that the low density scores are qualitatively in the right direction.

The number of steps between nodes (measured as average path length)[11] is another way to measure how *quickly* new information or orders could be passed through the network, but it can also tell us how efficient network organizations are at maximizing their relationship budgets. Networks with a shorter average path length and diameter should be able to respond more rapidly and efficiently to stimuli with the most efficient use of their time and resources. The average path length in all five of the networks we studied is less than three steps, with a maximum diameter of six steps. This means that all five networks are fairly efficient in terms of maximizing their secondary contacts. This is one area in which one possible "ideal" or target measure has already been established. The goal should be to have an average path length of just two steps. This is because research indicates not only that shorter paths are more important but also that any relationship based on more than three steps is most likely one in which a particular organization does not have access and certainly cannot influence (Friedkin, 1983, 1998; Burt, 1992). If node A is friends with node B, they are one step apart. Node A is two steps from node B's friend (node C), and node A is three steps from node C's friend (node D). Consider how difficult it would be for node A to exert influence on node D's behavior in that case. Noah Friedkin's study of six communications networks comprised of six different university departments demonstrated that there is a "horizon" of observability. This means that, beyond a certain number of steps (the distance between node A and node D, in this example), an individual is unlikely to know or be able to influence what someone else is doing. To illustrate this more concretely, let node A represent the director of communications for a city, node B is the assistant director, node C is the assistant director's executive assistant, and node D comprises a group meeting to which the executive (node A) is invited. Imagine that the director of communications (A) does not have the time to attend a meeting and instead sends her assistant director (node B) to participate and report back—the executive is two steps removed from the meeting (node D). Compare this situation to the following. The director (A) cannot attend the meeting and asks her assistant director (node B) to participate in her stead. The assistant director (B) is also busy that day, however, and so he asks his executive assistant (node C) to attend the meeting in his place. Not wanting the director to know that he did not attend the meeting in person, the assistant director (B) reports what happened in the meeting (D) to the director (A) based on the information provided to him by his executive assistant (C)—the director (A) is now three steps removed from the meeting, and the possibility of miscommunication or incomplete information multiplies accordingly.

In the five networks we studied, the longest path any node would have to take to reach another is six steps (five, averaged) measured as the *diameter* of the network. That is not too far, given the relatively large size of the networks, but it is still more than twice the ideal if the goal is an average of two steps (based on Friedkin's work). In the context of small-world/structural-hole theory (Burt, 1992) both two- and three-step relationships are important; a network with an average path length of just three steps is still a fairly tightly connected network.

Closeness centralization (normed) and average path length are different ways to get at a similar concept: How efficient are communications and coordination likely to be within the

[11] The number of steps between each possible pair of actors can also be measured without averaging the number of steps. The number of actors at various distances from each actor could provide us with information about the constraints and opportunities different actors have based on their position in the network and information about distance can also tell us which actors are the most costly to reach (Hanneman and Riddle, 2005, Ch. 7).

network? Average path length provides a more nuanced sense of how we might begin establishing specific goals for an ideal network.

Resiliency, redundancy, and single points of failure. Although a highly centralized network may be *efficient* in terms of communications and coordination, it is probably less *resilient* than a decentralized network that might have more built-in redundancy. Since all the sites we studied have moderately low closeness centrality scores, they might be more resilient to disruption. This assumes that redundant communications structures have developed to bridge the gaps between organizations that may not be formally connected but want to have access to each other's information. Alternatively, there might be multiple levels of coordination and communications, and we may not be capturing them all.[12] Information from the site visits suggests that this may be the case: Some emergency managers at military installations stated that tactical organizations, such as security forces and fire services, have their own distinct relationships with their civilian counterparts. The network surveys do reveal some of these relationships but require further analysis to parse out the potential tactical-operational distinctions.

A highly centralized network is less resilient because the most central node is a potential point of critical failure. Network centralization is thus inversely related to network resiliency. A network with more redundancy (multiple nodes tied to the same groups of partners) is more resilient to the removal or incapacitation of key network members. As mentioned in the "Most Influential Organizations" topic at the beginning of this section, those organizations with the highest betweenness centrality scores are also potential single points of failure in the network, since they are responsible for bridging the gaps between otherwise disconnected nodes. Their removal from the network leaves pairs of nodes completely disconnected from one another.

As a reminder, the organizations playing this broker role were similar across all five sites. The most between organization at three of the five sites (and, therefore, the organization that renders the broader emergency preparedness network most vulnerable) was a public health or medical organization. Public health/medical was also the third most between at the remaining two sites. The local emergency management office was the second most vulnerable organization in terms of betweenness at four of the five sites, and third most vulnerable at the fifth site. Since the local emergency management office has the greatest responsibility for preparing for and responding to emergencies, the vulnerability of this organization is problematic. The same is true of public health/medical organizations, which are responsible for a critical piece of the response puzzle. The vulnerability of each of these organizations suggests the need to create redundant ties for this organization and to establish a backup plan for the potential failure of these nodes.

Key Characteristics of Each Network

In Appendix G, we present the characteristics of the emergency management networks for each of the five networks in two ways: a figure depicting the network and a table of network statistics. In our discussion, we first identify key players in positions of influence or leadership

[12] We could better test this hypothesis if we were able to assess the number of cliques and subgroups and the degree of triadic closure within the network. This would allow us to assess the amount of redundancy in the network, as well as to better evaluate internal communications flows—identifying potential blockages or potential network efficiency in executing missions that require cooperation from the entire network—such as disaster response. The large amount of missing data in our current study precludes us from using these measures at this time, however.

by calculating measures of degree centrality and betweenness centrality for network nodes. We then look at the implications of the network's structure on communications flow, coordination, and innovation. We evaluate the flow of network communications and the potential for coordination (efficiency) and innovation (flexibility) within the network by calculating measures of normed closeness centralization, density, average path length, and average diameter. We then evaluate the network's resiliency, redundancy, and single points of failure, looking again at normed closeness centralization and density, and potential single points of failure using betweenness centrality measures.

Figure 5.1 is an example of a network diagram, which we use to depict each network in graphic form. This figure presents the key characteristics of the San Antonio metropolitan

Figure 5.1
Combined Military-Civilian Preparedness Network, San Antonio, Texas, Metropolitan Area

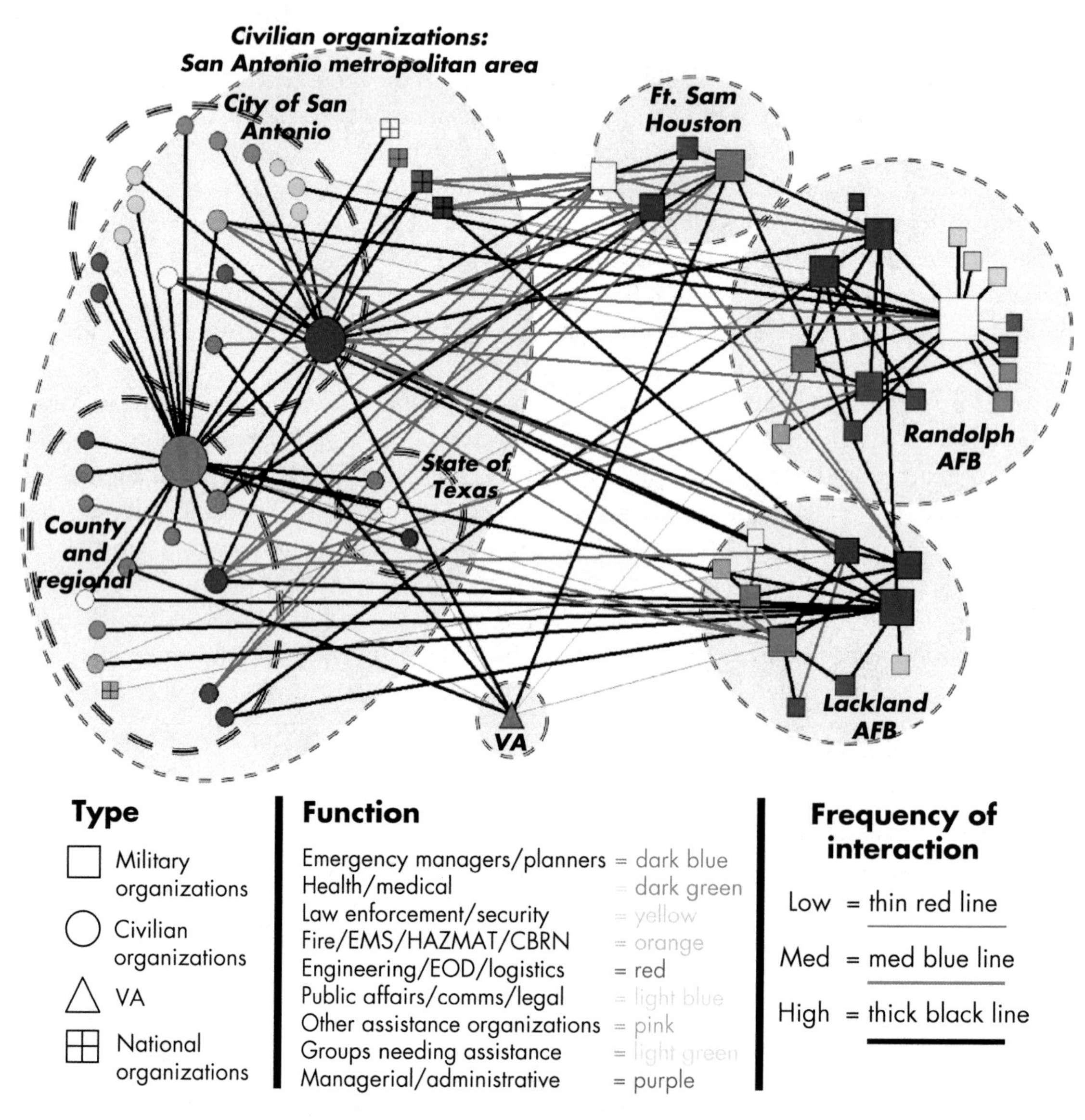

NOTE: EOD = explosive ordnance disposal. CBRN = chemical, biological, radiological, or nuclear.

RAND *TR764-5.1*

area network. The green clouds represent individual military installations, the VA, and civilian organizations, clustered by city, county/regional, or state organizations. Plain square nodes represent military respondents or organizations—all nodes within the installation clouds are square. Round nodes and hatched square nodes represent civilian respondents or organizations and local representatives of national organizations, respectively. The colors refer to functions. For example, blue nodes represent emergency management functions, and green nodes represent health and medical functions. The size of the nodes represents their degree centrality.[13] Larger nodes have higher degree centrality scores and, hence, are more influential within the network.[14]

Figure 5.1 highlights the following key characteristics of the San Antonio metropolitan area network:

- The San Antonio civilian emergency manager (largest dark blue circle) and public health/medical (largest green circle) nodes—located at the center of the "Civilian organizations" cloud—are the largest, indicating that they are more influential than all of the other surrounding nodes.
- The Randolph AFB law enforcement/security node (largest yellow square) is central to the installation's internal network and interacts frequently with counterparts across functions both within the installation and in the civilian community.
- The emergency management/planning (dark blue squares) and medical (green squares) nodes at Randolph AFB, Lackland AFB, and Fort Sam Houston are equally influential within the network and more influential than most of their surrounding nodes.
- At Fort Sam Houston, the emergency management/planning, health/medical, and law enforcement/security (yellow square) nodes are almost equally important to the network.
- Although the most central organizations within each neighborhood tend to connect with like functions across green clouds, their connections are not limited to similar functions. Ties are still denser within each cloud than across clouds, however.

Figure 5.2 summarizes key characteristics of the Norfolk and Virginia Beach network:

- The Norfolk and Virginia Beach civilian emergency managers (largest dark blue circles), Virginia Beach Department of Public Health (largest green circle), and the Virginia Beach Fire Department (largest orange circle) are the largest civilian nodes, indicating that they are the most influential civilian nodes.
- NS Norfolk emergency management office (largest dark blue square) and NMC Portsmouth (largest dark green circle) are similarly influential to the Virginia Beach civilian emergency management office (the larger of the two blue circles).
- NS Norfolk Fire Department (large orange square) is similarly influential to the Virginia Beach Fire Department (largest orange circle) and to the Norfolk civilian emer-

[13] Degree centrality is measured in terms of the proportion of total possible dyadic connections that each network member has.

[14] Although we tried to select survey respondents who were more likely to play an important role in the larger local disaster preparedness network (from both military and civilian perspectives), and although multiple respondents sometimes nominated other network members, the significance of respondent centrality is partly mitigated by the fact that respondents themselves were responsible for defining their network neighborhoods.

Figure 5.2
Combined Military-Civilian Preparedness Network, Norfolk/Virginia Beach, Virginia, Metropolitan Area

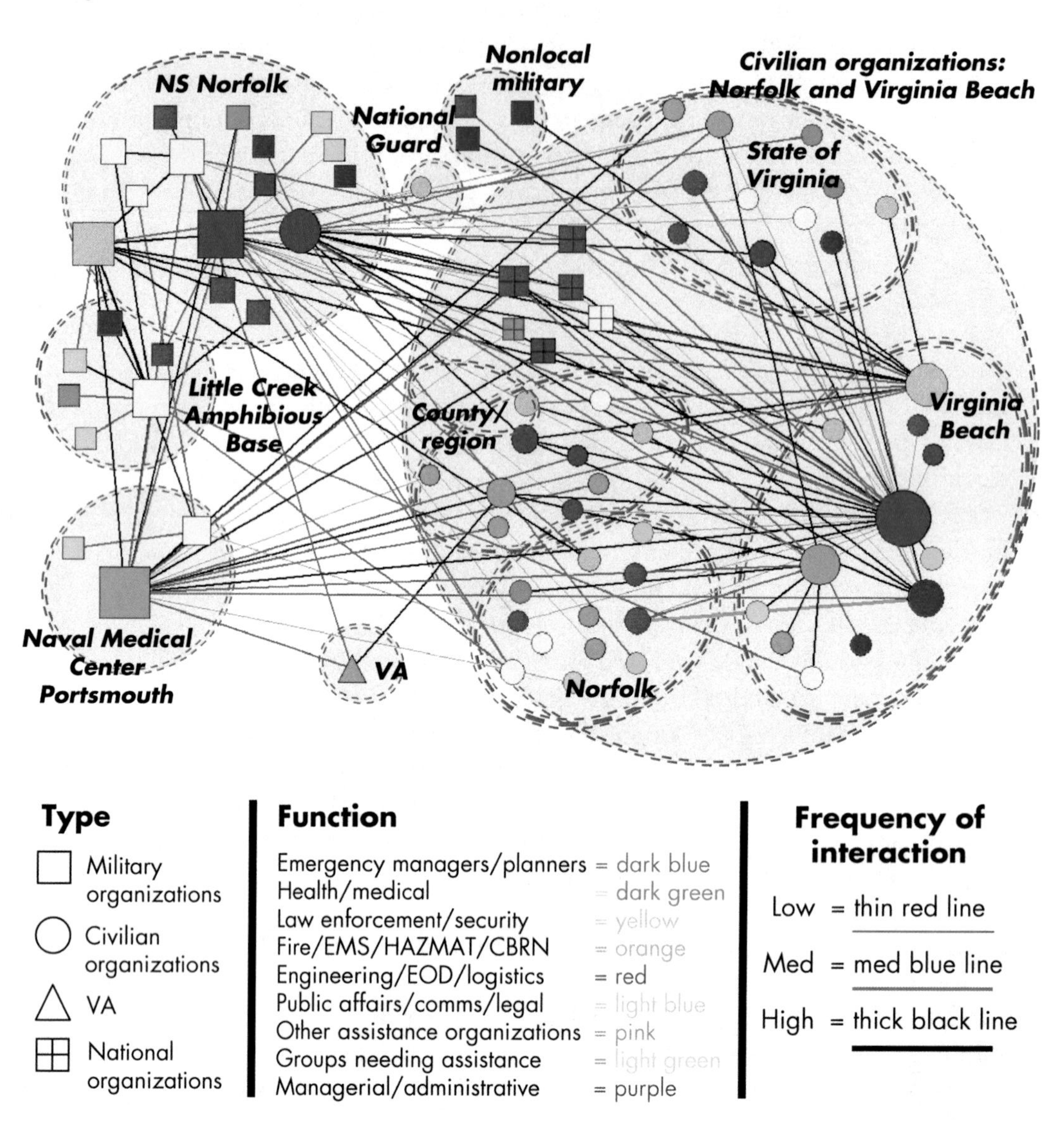

RAND *TR764-5.2*

gency management office (smaller of the two large blue circles) and Virginia Beach Public Health (largest green circle).

- Each of the most influential nodes is connected to at least one, but usually more, of the other most influential nodes, which helps to bridge the divide across both civilian and military and differing functional communities.
- The densest inter-organizational ties are in the public health/medical community (green squares, circles, and triangle). The star-shaped neighborhoods are connected to each other via multiple ties.

Figure 5.3 summarizes key characteristics of the City of Columbus and Muscogee County network:

Figure 5.3
Combined Military-Civilian Preparedness Network, City of Columbus and Muscogee County, Georgia

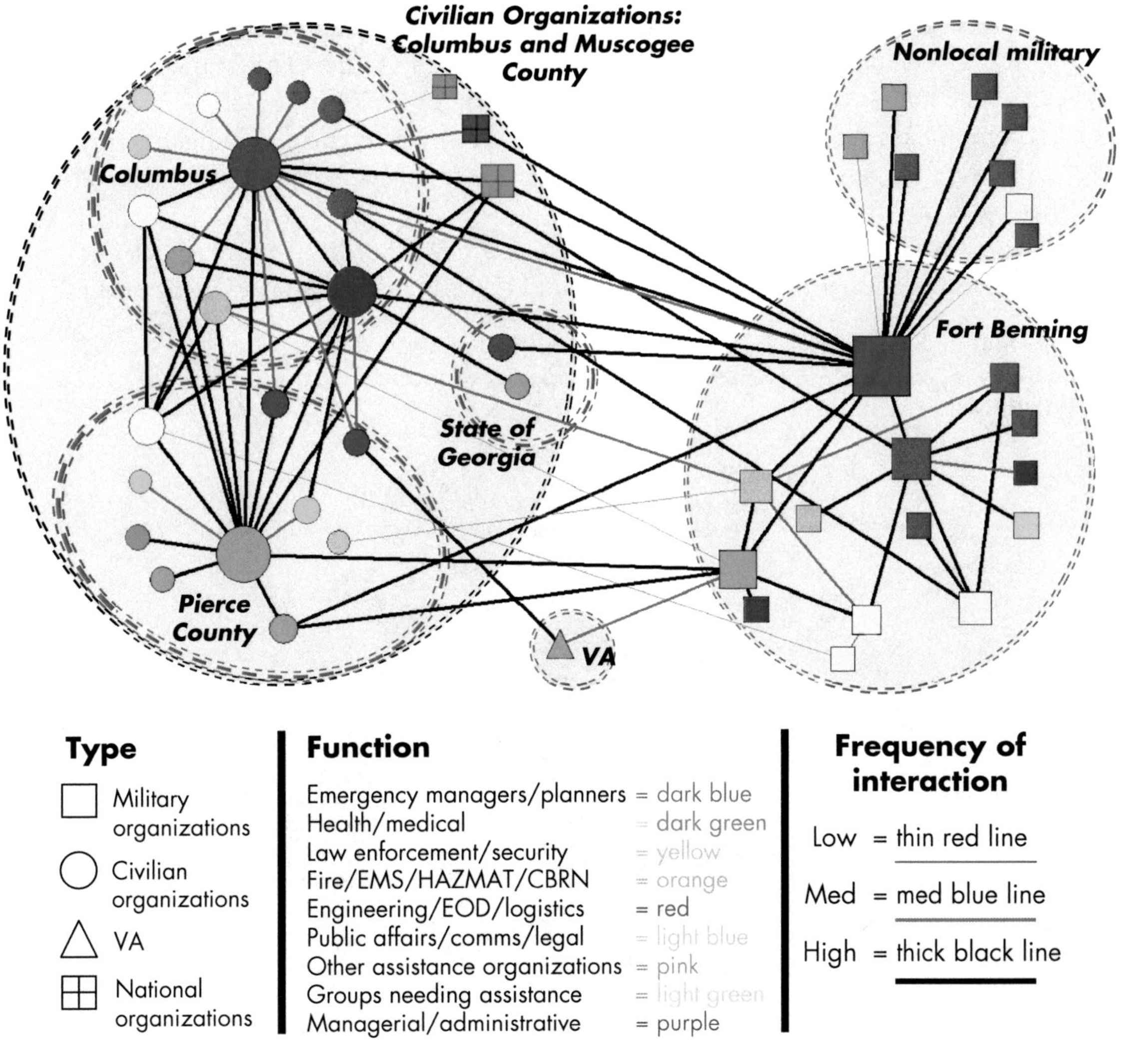

RAND *TR764-5.3*

- The Fort Benning emergency management office (largest dark blue square), the Muscogee County Department of Public Health (largest green circle), Columbus's emergency management office, and the Office of Homeland Security at City Hall (large purple circle) are the most influential nodes in the network.
- The Fort Benning emergency management office (largest dark blue square) is also the military organization with the largest number of ties to civilian organizations.
- The VA (green triangle) provides a bridging tie between Martin Army Community Hospital (largest green square) and emergency management offices from other Georgia counties (small blue circle directly connected to the VA).
- MMRS (small green circle at the bottom of the civilian cloud) provides redundant ties between public health (largest green circle) and Fort Benning emergency management and between public health and Martin Army Community Hospital—establishing some resiliency across the broader network.

Figure 5.4 summarizes key characteristics of the City of Tacoma and Pierce County network:

- The most influential nodes in the network are the Pierce County emergency management office (largest dark blue circle), the McChord Planning Office (largest dark blue square), and McChord's Medical Flight (largest green square).
- Tacoma/Pierce County Health Department (largest green circle), Madigan Army Medical Center (largest green in the Fort Lewis cloud), and McChord's antiterrorism office are also very influential.
- The Tacoma/Pierce County network seems to have the densest pattern of ties across neighborhoods that we have seen in any of our cases. This density is apparent based on

Figure 5.4
Combined Military-Civilian Preparedness Network, Cty of Tacoma and Pierce County, Washington

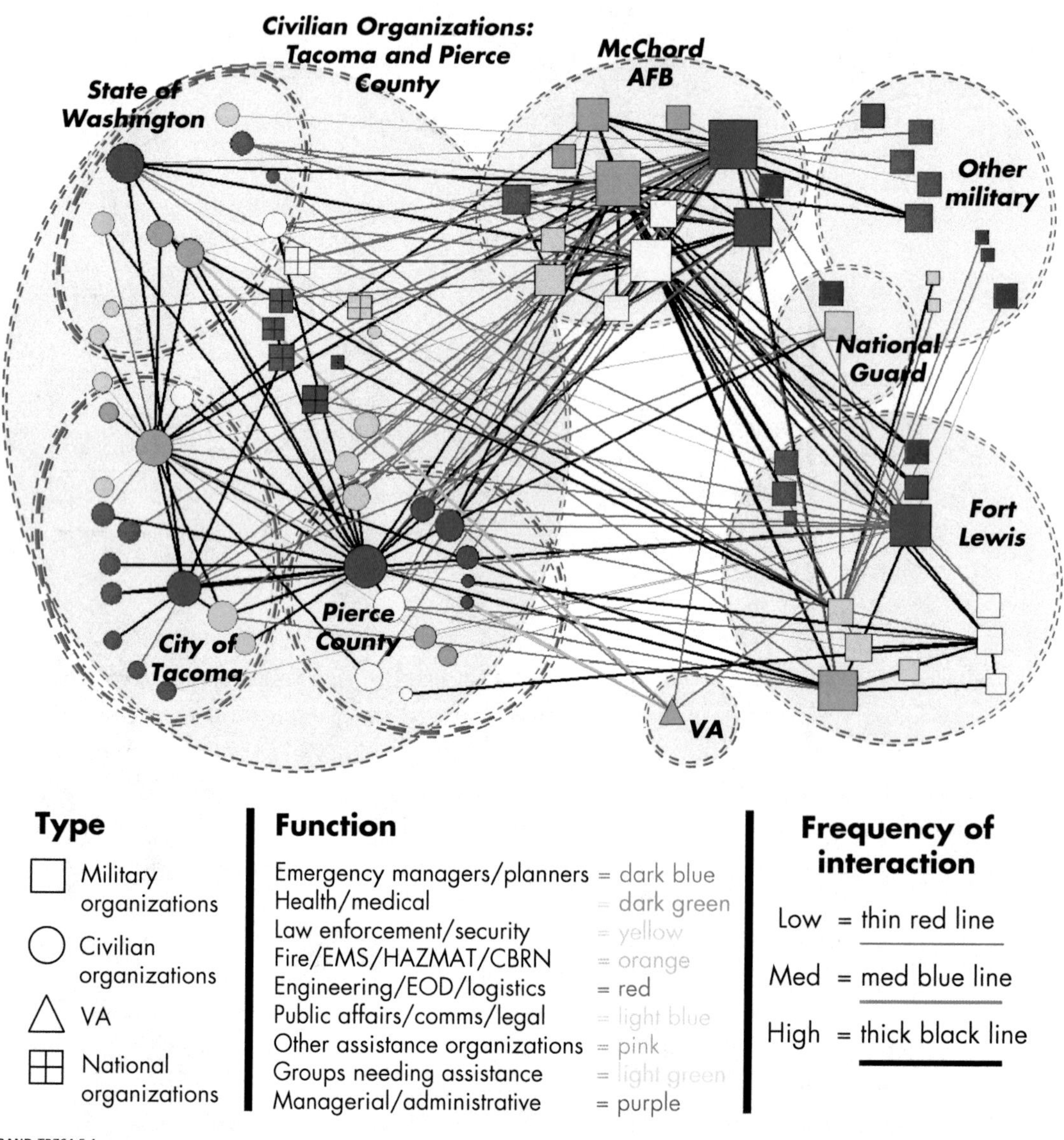

RAND *TR764-5.4*

the figure but is not well reflected in the statistics (shown in Table G.7 in Appendix G) because of the skewing effect of so many pendant nodes.

- Survey respondents nominated more nonlocal military organizations as being important to their emergency preparedness networks than in any other case. These are the blue and red squares located in the "Other military" cloud and correspond to U.S. Northern Command (blue square), in addition to U.S. Army North, U.S. Air Force Civil Engineer Support Agency, U.S. Air Force Mobility Command, and the Air Staff's Readiness and Emergency Management Office. We believe these nominations to be largely a function of McChord's active role in ongoing overseas conflicts.

Figure 5.5 summarizes key characteristics of the City of Las Vegas and Clark County network:

Figure 5.5
Combined Military-Civilian Preparedness Network, City of Las Vegas and Clark County, Nevada

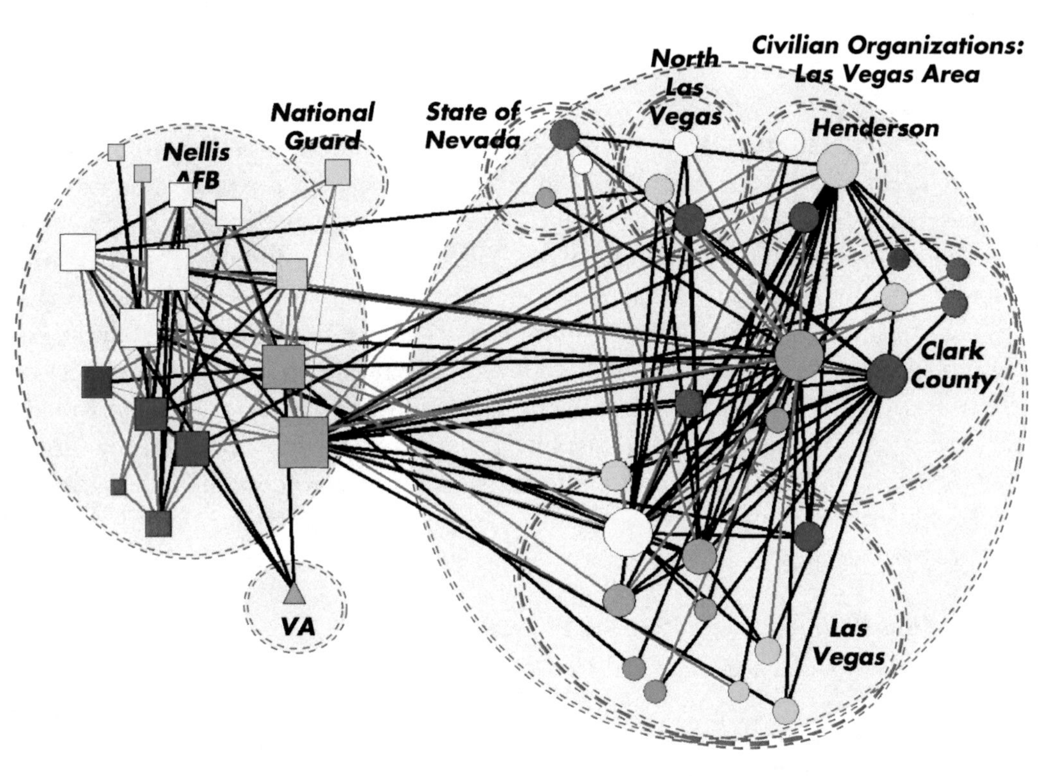

RAND *TR764-5.5*

- There are a number of similarly sized—and therefore influential—nodes in the network, including the Bioenvironmental Engineering Flight in the 99th Aerospace Medicine Squadron Nellis AFB (largest dark green square), the MOFH at Nellis AFB (second largest dark green square), Southern Nevada Public Health (largest dark green circle), the Las Vegas Police Department (largest yellow circle), the Clark County Office of Emergency Management (largest dark blue square), and the Henderson Fire Department (largest orange circle).
- The Nellis AFB network is an extremely dense network—perhaps the densest neighborhood we have seen.
- The VA is connected to the Nellis AFB network but is completely disconnected from the civilian community.

Summary of Findings

The following key findings emerged from our analysis of the social network surveys, by site and across sites.

Most Influential Organizations

Across all five sites, and in both the civilian and military communities, the most influential organizations, based on network structure in general and specifically on degree centrality or betweenness centrality scores, were consistently emergency management and planning, health and medical, and security and law enforcement offices, followed by fire services. These findings align closely with findings from the site visits described in Chapter Four about which organizations tend to be involved in the civilian planning process. Also in line with our findings from Chapter Four, the more expansive view of community expressed by civilian interviewees was illustrated by the fact that organizations identified by civilian survey respondents tended to include local organizations as well as those of neighboring cities, counties, and sometimes even states. Military respondents tended to define their community somewhat more narrowly but still identified local emergency management, public health, law enforcement, and fire organizations as key partners.

Communications Flow, Coordination, and Innovation

Network analysis provides a more global view of the civilian and military networks that hopefully complements, but also expands on, the information we obtained from our site interviews. So, while it may be true, as we concluded in Chapter Four, that fire services and public health organizations seem to be more connected across military-civilian boundaries, emergency management and law enforcement/security organizations also stand out as particularly influential players that help improve coordination across each of the larger site-specific networks. These organizations are important not only for increasing coordination across civilian and military communities but also for connecting otherwise disconnected organizations within either the civilian or the military community. Our network-analysis results, consistent with conclusions gleaned from site visit interviews, indicate that there is relatively little interaction between military planners and the local VA facilities and little interaction with the National Guard (except for the Guard's Civil Support Teams, which were mentioned as key members of the emergency preparedness network at three of our five case-study sites).

Overall, our data indicate that the networks at all five sites are fairly decentralized and not very densely connected, which means that communications and coordination across the networks are probably less efficient than would be the case in more centrally managed or more densely connected networks. However, we recognize that the apparent decentralization of these networks may be an artifact of our survey methodology. Our data do indicate that communications tend to be stovepiped within functional communities in each network, such as the public health/medical or law enforcement/security community, and within a local military or civilian community. Despite this stovepiping, there is a fair amount of regular contact across communities and across functions at various levels; it is just not as common or as frequent as it is within functions and communities.

Resiliency, Redundancy, and Single Points of Failure

A highly centralized network is a less resilient network because the most central node, or organization, is a potential point of critical failure. Network centralization is thus inversely related to network resiliency. A network with more redundancy is more resilient to the removal or incapacitation of key network members.

Each of our five sites is fairly decentralized and, as such, might be more resilient to disruption. There may also be redundancies at multiple levels of coordination—for example, organizations may have both tactical and operational relationships maintained by different individuals or offices. Information emerging from the site visits suggests that such redundancy exists: Several installation emergency managers noted that the tactical organizations, such as security forces and fire services, have their own distinct relationships with their civilian counterparts. Our network surveys also reveal some of these relationships.

Those organizations that act as brokers between otherwise unconnected pairs of organizations are also potential single points of failure in the network, since they are responsible for bridging the gaps between these otherwise disconnected nodes. Their removal from the network would leave some nodes completely disconnected from one another and, potentially, from the broader network entirely. The organizations playing this broker role were similar across all five sites. The organization most often fulfilling the broker role at three of our five sites (and therefore, the organization that renders the broader emergency preparedness network most vulnerable at that site) was a public health or medical organization. The second most common broker at four of our five sites was the local emergency management office.

Implications for the RAND Planning Support Tool

There is still no perfect way to measure "preparedness" a priori. However, SNA can help elucidate the structural and relational factors that either facilitate or hinder cooperative preparedness planning—particularly if we compare findings across the five networks we studied. Verifying the SNA implications of our findings (e.g., which networks are better at communicating, which agencies are more influential) with actual data would be an area ripe for research. Such research would leverage our initial SNA results to both improve data-collection methods and develop more accurate measures of resiliency, redundancy, efficiency, and flexibility relevant to local disaster preparedness planning. It would be important to ensure that identified measures correlate specifically with success in emergency management–related networks, narrowing the focus from the broader organizational findings on which our hypotheses are based.

As a first step, we would need to obtain more complete information about each of the local preparedness networks, whether through additional surveys, interviews, or information from event or exercise reports. The best place to start would be with those organizations that turned out to be the most commonly nominated within each community. The fact that certain organizations came up repeatedly in completely separate surveys suggests that they are, at the very least, highly visible within the local disaster preparedness community.

A potential SNA component of the planning tool would enable the user to acquire and evaluate his or her network statistics for density, centrality, betweenness, and efficiency and identify opportunities for better local networking. Background information would provide a discussion of the advantages and disadvantages of a dense versus a more dispersed network, for example, and the trade-offs required to increase the number of ties within a local disaster preparedness community. SNA has promising possibilities for helping local emergency management players to improve planning and coordination by assessing the strength, breadth, and resilience of their local networks. This capacity is not currently available without developing expertise in one of the existing network-analysis software packages, which are not specific to emergency management networks. Anecdotally, many of our interviewees proclaimed that well-developed social networks are the foundation of successful local emergency management planning. This is true in terms of communications speed, for example, but the complexity of some networks might actually render efficient coordination more difficult. An SNA tool could help establish the parameters of a "well-developed" network and enable the user to measure his or her own network against these parameters—scaled to population density, geographic factors, and the time-geographic availability of resources that would augment a local response.

Such a tool might establish one or more "ideal" emergency management networks against which users could measure their own networks. There are two possible methods for establishing an ideal network model: (1) top down and (2) bottom up. A top-down build would look something like Kapucu's (2005) model. Kapucu used responsibilities designated under the FRP (superseded by the National Response Plan and now the NRF) to establish the ideal network parameters. He compared this network with actual response performance as reported in post-9/11 FEMA situation reports. We could construct an ideal network based on federal or state guidance or on a combination of the two. This model could then be compared to specific local response networks—such as those represented by our survey respondents. A comparison between the "ideal" and the "real" would allow identification of the following:

- missing relationships in certain networks or for certain event types
- redundant relationships (including the advantages and disadvantages of that redundancy)
- potential single points of failure
- relationships between key players where path lengths are too long to be useful (e.g., everyone should be connected to important players A–C by no fewer than two lengths, but some organizations might be five lengths from these key organizations)
- organizations with which they should be interacting (e.g., planning, exercising, responding) based on either the user's organizational function or the type of event.

A bottom-up approach would look much like our analytic efforts in this chapter. Here, we established the qualities of a desirable emergency preparedness network (resiliency, redundancy, efficiency, and flexibility) and then identified corresponding network measures. Some of these measures entail trade-offs; there is a trade-off between redundancy (which allows a network to be more resilient) and efficiency (which allows a network to respond more rapidly to new inputs). Any useful SNA tool would have to include discussions of trade-offs such as these and should also provide for network visualization: The ability to "see" one's network of relationships and responsibilities can often provide insights into ways to improve coordination, communications, or resiliency of the network.

A somewhat simple and simplistic starting point for further development of SNA in the context of local disaster preparedness might be to help compile a functional list of organizations relevant to local emergency planning, to which names and their contact information could be attached and presented as a "network-looking" graphic. Further research would be needed to proceed with the development of a full-scale SNA tool. Next steps along this path would be to (1) collect more systematic and more complete information about networking connections among local agencies, so that a network can be represented in graphic and even statistical terms; (b) develop the criteria and visual and statistical representation of "ideal" networks; and then (c) develop and field test a tool that could be used to represent local networks and suggest steps toward achieving the structure of an "ideal" network.

CHAPTER SIX

Framework for a Local Planning Support Tool

The ultimate goal of this project is to facilitate local risk-informed capabilities-based planning by military installations and civilian agencies (including local VA providers) in communities across the United States. We aim to create a planning support tool that will enhance cooperative local planning and other preparedness activities to mitigate risk from major disasters.

In preceding chapters, we have described civilian and military policies and programs underlying local disaster preparedness (Chapter Two) and risk management and capabilities-based planning concepts (Chapter Three). These all constitute top-down policy with which bottom-up local planning—our focus—must be consistent. We have also summarized findings from our interviews at five sites across the country by describing how local civilian agencies and military installations currently plan for and respond to major disasters (Chapter Four) and the nature and extent of network ties among them (Chapter Five), both of these to further establish the basis for a local planning support tool that will meet the needs of users.

In this chapter, we describe the final elements of the formative research conducted to guide development of a new local planning support tool. During our site visits, we garnered information about what civilian and military emergency management personnel would find most useful to support their disaster preparedness efforts—broadly exploring their needs related to risk assessment, planning, and event management. Through a separate review of websites and documents and complemented by our site interviews, we also inventoried existing preparedness-oriented support tools, also classified into these three categories. In the discussion in this chapter, we draw on this information to highlight the desired characteristics of a new planning tool, describe the planning framework within which a tool would be developed, and provide an example of the tool itself.

Perceived Needs: Desired Features of a New Planning Support Tool

In our interviews at the five study sites, we asked respondents to describe not only their current local emergency preparedness activities and tools but also the desired characteristics of a potential new RAND planning support tool. The functional characteristics, implementation features, and desired capabilities they wanted are discussed in this section.

Functional Characteristics

- Integration of tools: Some interviewees pondered the relative merit of simply increasing coverage with existing adequate tools, versus integrating current tools, versus creating a new tool. One suggested a "one-stop shopping" approach to help synchronize different tools and databases. Our interviews strongly suggested that local emergency planners are not aware of the full range of tools in use, including which tools are available to them and how these tools complement or supplement each other.

- Applicability to specific preparedness and response phases: Comments on the potential scope and scale of a new tool were wide ranging. Interviewees expressed desire for a tool to better support planning, real-time event management, or both. A new tool might have a single or multiple purposes, including planning, operations, logistics, budget, risk assessment, capabilities-based planning, teaching or refresher training, exercises, or facilitation of requests for assets from local partners (e.g., civilian requests for military support).
- Scope: Our interviewees thought that a tool should be flexible enough to accommodate different emergency situations or scenarios. It could be local or nationwide, serve as simply an information repository or as a decision support tool, and capture activities and assets either by function or as a simple enumeration absent function. A tool should be relevant from local to more central levels, potentially with different content and level of detail for different levels of disaster management. Some individuals commented that a new tool should be scalable from purely local to increasingly aggregated localities as needed for planning or response purposes.

Implementation Features

- Ease of use: According to interviewees, a new preparedness support tool should be well accepted and user friendly (e.g., "plug and play") with non-burdensome user requirements. It should be used regularly enough to be familiar to all, easily updatable, and easy to maintain.
- Access: A new tool should be available to those who need to use it but protected from those who do not—e.g., a password-protected shared-access site that is compatible across firewalls and a system that is interoperable across actors. Some interviewees suggested a "pull" mechanism, and others suggested a "push" mechanism for accessing a jointly used tool or website.
- Portability: Several individuals noted the importance of portable (e.g., handheld) wireless capabilities (e.g., cell phone or laptop), and they described potential event-based limitations in these (e.g., use at the EOC versus on site, difficulty of using wireless equipment on military installations).
- Security: Military interviewees noted that military emergency planning often includes a classified component (or at least is housed on a classified system), necessarily limiting information sharing with civilian counterparts. Civilian planners also shared security concerns about their data and argued for security measures, including password protection of information and tiers of access, so that only a small number of individuals had access to data about organizations and assets.

Desired Capabilities

- Risk assessment
  - automated (or standardized and more efficient) risk assessment process, rolling up assessments of threats, vulnerabilities, and consequences

- Planning
  - automation of good guidelines (e.g., the VA hazard vulnerability assessment, DoD Joint Antiterrorism Guideline
  - integration of source documents across agencies and actors, including those that outline requirements or functional responsibilities, as well as AARs and best practice–related documents
  - updated lists of key actors across major agencies and their contact information
  - definition of the roles and responsibilities of different local actors
  - generation of operational checklists (e.g., task-completion checklist, job action sheets by role/function, such as an electronic version of tactical worksheet)
  - support to contingency planning
  - matching of capabilities to needs
  - resource typing and inventories (e.g., local personnel and their skills and capabilities; reconciliation of civilian and military resources and assets, even if labeled differently)
  - generation of options and resources needed to respond to different types of events
  - prioritization of SOPs
  - incorporation of gaps identified in exercises into updated local plans
  - information warehousing (e.g., structural integrity of buildings, SOPs across agencies, emergency management library, and AARs and lessons learned).
- Event management
  - geographic information system (GIS) mapping (e.g., event, assets) and portable plotting capabilities
  - communications: Interoperable emergency management communications across key local civilian and military agencies, such as videoconferencing, and rapid communications modality, such as electronic chat capabilities
  - situational awareness: facilitation of a common operating picture for all key actors; management of the surge in telephone calls during an event; more fruitful use of real-time disease information, tracking patients as they move across sites or facilities, including across military, VA, and other local civilian facilities
  - asset visibility: shared visibility of assets and inventory, shared database of credentialed personnel during an event, operational checklists (e.g., task completion), and resource and capabilities tracking
  - marshalling available resources or assets by a process that is more efficient than current modalities (e.g., by hand or via telephone).

Existing Preparedness Support Tools and Resources

Through Web-based searches, review of documents, and site interviews, the RAND team identified a number of tools and resources currently available and used to support local activities. We have characterized an inventory of approximately 30 of these according to such factors as functional support area (risk assessment, planning, event management), access (public versus commercial), hazard(s) addressed, outputs, required user inputs, and target audience (Appendix H; searchable version available from the authors upon request). Information about functional support area, access, scope in terms of hazards addressed, and target audience allows readers to identify other tools that may meet their specific needs and whether and how they might access such tools; information about inputs and outputs provide a base of information for the RAND team in considering approaches to the new planning support tool.

The NPG call for certain activities to be conducted in the "preparedness cycle in advance of an incident" at all levels of government. The cycle of preparedness for prevention, protection, response, and recovery missions includes the following elements:

- plan
- organize and staff
- equip
- train
- exercise, evaluate, and improve.

Recognizing that each of these is a critical activity that must be accomplished, we have, as previously noted, focused our efforts in this project on the "plan" function.

In this section, we describe some of the tools that appear to be broadly applicable to risk assessment, planning, and event management, including tools that were frequently mentioned by our respondents as used or useful. We also include tools that we have identified as important, even if not mentioned in site visit interviews, if they reflect an area that we should not attempt to replicate or if they may be complementary or supplementary to the proposed RAND tool. We use the broad areas shown in Figure 6.1 to organize our descriptions. The risk assessment stage in Figure 6.1 corresponds to the process presented in Figure 3.2 in Chapter Three. The planning box in Figure 6.1 suggests operational steps within the context of capabilities-based planning, which was introduced in Chapter Three. The event management box was introduced in that chapter and was organized based mainly through synthesis of information from interviews. The exercises and lessons-learned boxes were derived from analysis of national planning guidance. The community networking box reflects our conceptualization of where the SNA described in Chapter Five fits into the risk-informed planning process.

Risk assessment as a preparedness planning activity includes threat assessment and vulnerability and consequence assessment, which lead to a final activity, producing a set of priority risks. The set of priority risks feeds information into the planning phase, especially capabilities-based planning, which includes identifying capabilities and resources needed and resource gap analysis based on comparing resources needed and available. In addition to the inputs from the risk assessment, other planning activities may also be enabled or informed by community networking, exercises, and lessons learned from actual incidents. Planning in turn feeds into event management, including such activities as communications, situational awareness, and asset visibility during an emergency event.

Risk Assessment

A number of organizations have developed tools to facilitate completion of one or more aspects of a risk assessment. The largest number of tools we identified have been in the area of vulnerability and consequence assessments (Appendix H).

Vulnerability Assessment and Consequence Assessment. We identified both civilian and military tools that support vulnerability assessments. The most frequently mentioned and used civilian tools were those developed by Kaiser Permanente and the VA. DoD has developed a number of tools for conducting vulnerability assessments. At the installation level, the most commonly mentioned was the JSIVA.

Figure 6.1
Selected Preparedness Activities

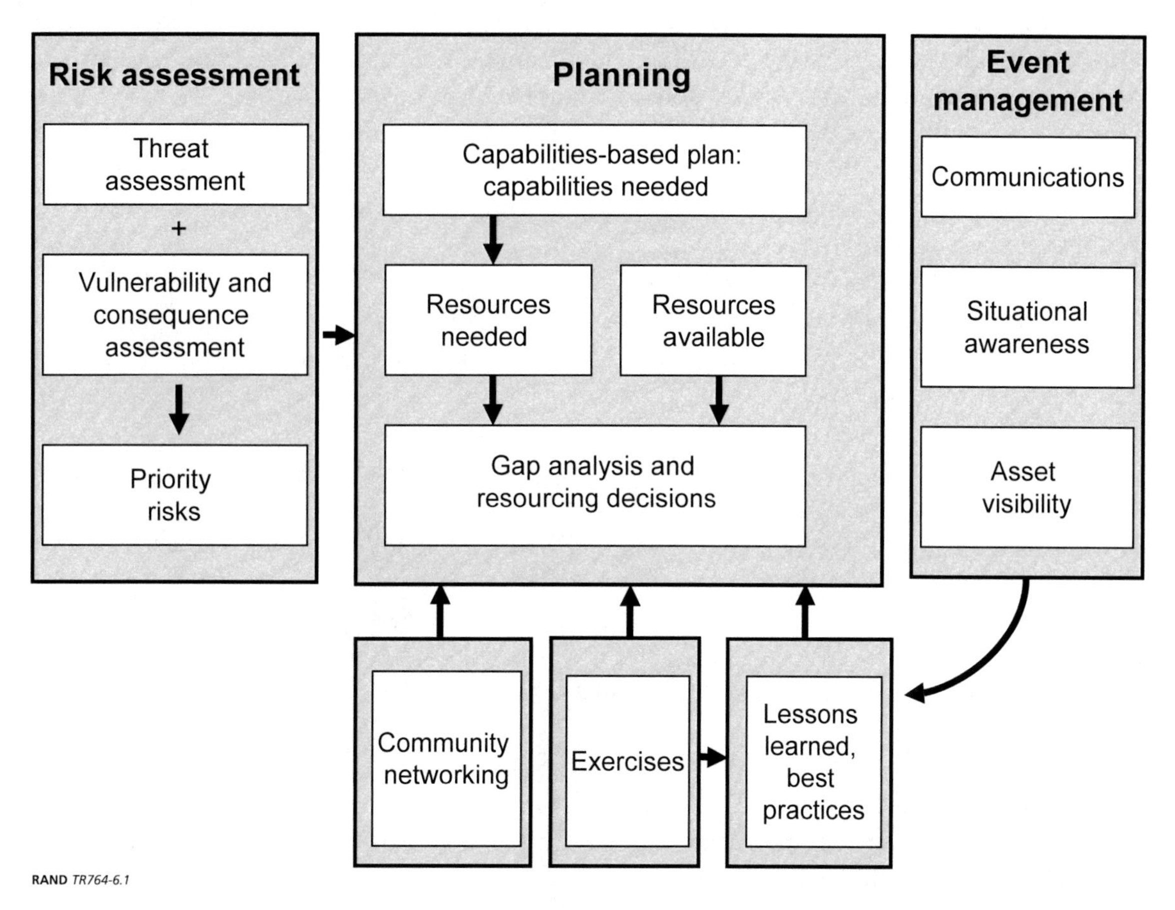

RAND *TR764-6.1*

Kaiser Permanente Hazard Vulnerability Analysis Tool. Kaiser Permanente's Hazard Vulnerability Analysis assessment tool is a publicly available worksheet designed for hospitals and other medical facilities to calculate the risk to a facility in the case of naturally occurring events, technologic events, manmade events, and events involving hazardous materials. On-site users fill out the worksheet with the likelihood that future events may occur (threat) and their impact (vulnerability and consequences) and presence of mitigating factors to obtain a risk score (calculated based on probability and severity, with the latter comprised of impact and mitigating factors) for each type of hazard, which can then be used to guide planning and exercises (Kaiser Permanente, 2001).

***VA 2008* Emergency Management Program Guidebook.** The VA 2008 *Emergency Management Program Guidebook*, directed at both regional VISN offices and local treatment facilities, outlines the compliance standards for emergency management in six critical areas: communications, resources and assets, safety and security, staff responsibilities, utility management, and patient clinical and support activities. It also lays out a nine-step process for the development, maintenance, and evaluation of an emergency management program. This process takes up the vast majority of the guidebook, which devotes individual sections to each step (VA, 2002).

JSIVA. JSIVA is a DoD antiterrorism/force protection assessment. JSIVA is used to determine an installation's vulnerability to mass casualty terrorist attack, and it provides procedural and technical options to mitigate the risk to installation personnel. JSIVA is designed for all DoD installations with at least 300 personnel (DTRA, undated).

Hazards U.S. Multi-Hazard (HAZUS-MH). HAZUS-MH is a risk assessment software program developed by FEMA and designed to analyze the potential losses from natural disasters—floods, hurricane winds, and earthquakes. HAZUS-MH draws on past data to provide probability of occurrence (threat) and uses GIS software to map the consequences and damage caused by floods, hurricane winds, and earthquakes, as well as economic losses due to damage to buildings and other infrastructure. The program further estimates the impact on the local populations (FEMA, 2009b).

Electronic Mass Casualty Assessment and Planning Scenarios (EMCAPS). EMCAPS is a public software program developed by Johns Hopkins Office of Critical Event Preparedness and Response to model mass casualties resulting from various disaster scenarios. Specifically, EMCAPS addresses chemical, biological, radiological, and IED attacks. Users can input scenario characteristics, such as bomb size and population density, and the model will output casualty estimates. Models are available for plague, food contamination, chemical (blister, nerve and toxic agents), radiological, or explosive attacks.

Consequence Assessment Tool Sets (CATS). Developed by the U.S. Defense Threat Reduction Agency, FEMA, and Science Applications International Corporation, CATS is a suite of hazard models to calculate the casualties and damages of natural disasters, terrorist incidents, and industrial accidents. The software also generates resource requirements to mitigate damages and deal with the aftermath (SAIC, 2009).

Planning

Capabilities-Based Planning. In response to HSPD-8, DHS established the parameters of a capabilities-based planning approach to disaster and emergency preparedness planning. This approach includes the following tasks:

- Identify a plausible range of major events.
- Identify the prevention, protection, response, and recovery tasks that require a coordinated national effort.
- Define risk-based capabilities and levels of capabilities that minimize the impact on lives, property, and the economy.

As part of the approach, DHS has published and is updating several products, including the National Planning Scenarios, the UTL, and the TCL.

Sync Matrix. Sync Matrix is an interactive decision support tool designed by Argonne National Laboratory to provide a structured approach to developing local preparedness plans. The software application provides a method for developing plans for a variety of emergency and disaster response scenarios. It also provides a platform—a collaborative workspace (via the Microsoft® .NET Framework)—for communications across agencies and as a record of planning processes. Sync Matrix is available to state and local governments through FEMA's Preparedness and Program Management Technical Assistance programs (ANL, undated [b]).

Capability Based Planning Methodology and Tool (CBMPT). The CBMPT, developed by Johns Hopkins University Applied Physics Laboratory, is designed to map local needs during an emergency to the public safety organization responsible for preparing and responding to those needs. The CBMPT also tracks the operational capabilities required of the emergency responders and areas in which capabilities need improvement. The CBMPT can be applied to all types of hazards. (Note: As of this writing, commercial availability of the CBMPT is still pending.)

Preparedness for Chemical, Biological, Radiological, Nuclear, and Explosive Events: Questionnaire for Health Care Facilities. This questionnaire for health-care facilities, developed by AHRQ within HHS, is meant to assess the readiness of health-care facilities (both individual and multihospital systems) for CBRNE events in terms of administration and planning; education and training; communications and notification; patient (surge) capacity; staffing and support; isolation and decontamination; supplies, pharmaceuticals, and laboratory support; and surveillance. The questionnaire is designed to be analyzed on site to gauge site readiness in terms of these capabilities.

Resources Needed

NIMS Incident Resource Inventory System (NIMS-IRIS). NIMS-IRIS is a FEMA resource inventory database designed for input by communities of their 120 typed resources (i.e., based on FEMA's resource typing scheme). The database is intended to help inventory and identify resources available for emergency response operations based on mission requirements, capabilities of resources, and response time (FEMA, 2007c).

Emergency Preparedness Resource Inventory (EPRI). EPRI is a public tool developed by AHRQ to facilitate regional planners' ability to create inventories of critical resources required to respond to bioterrorist attacks. The tool can be regionally customized and is Web-based to allow multiple users to enter inventories. The tool also provides automated reports to use in exercises and disaster response.

Exercises

Homeland Security Exercise and Evaluation Program (HSEEP). HSEEP provides a standardized methodology and terminology for exercise design, development, execution, evaluation, and improvement. The HSEEP is intended to be a national standard for all civilian emergency exercises, providing consistency and unity of exercises at all levels of civilian government (DHS, undated [a]).

Lessons Learned and Best Practices

Lessons Learned Information Sharing (LLIS) System. LLIS, developed by DHS, is a national, Web-based forum intended to enable emergency managers to share lessons learned and best practices. The forum is open to all emergency managers and those associated with emergency management and response (DHS, undated [b]).

Responder Knowledge Base (RKB). RKB is a FEMA-sponsored database of information on products, standards, certifications, and grants. It is intended for emergency responders, purchasers, and planners.

Event Management

WebEOC. Developed by software firm ESi®, WebEOC is a commercial, Web-based all-hazards tool that supports event management through communications, situational awareness, and asset visibility. Its features include but are not limited to a status board, chat capabilities, checklists, contact information, report capabilities, National Weather Service (NWS) alerts, and a file library for plans. Together, these features provide situational awareness and real-time management of information and assets (ESi Acquisition, undated).

Tools Applicable to More Than One of the Phases

LiveProcess. LiveProcess is a commercially available tool, developed by the LiveProcess company. Like WebEOC, it was designed for all-hazards emergency planning and management in hospitals and other medical facilities. LiveProcess includes hazard-vulnerability analysis capabilities, an incident command system, a NIMS compliance tool, policy management, and a reference library. Further, LiveProcess is designed to help identify threats, standardize plans and training, measure competency, and provide communications capabilities (LiveProcess, undated).

Previstar™ Continual Preparedness System (CPS). Previstar CPS is a commercial, automated system of processes for preparedness, response, and recovery, based on the NIMS framework. The system includes a planning generator, exercises, task lists, status boards, communications capabilities, inventories, needs, deployments, costs, damage assessments, and modeling capabilities. Overall, Previstar CPS is meant to enable the management of multiple complex systems, including resources, personnel, and operational processes (Previstar, undated).

Opportunities to Bridge Gaps in Local Preparedness

Existing tools for capabilities-based planning tend to be linked to specific threats or specific localities. DHS has also developed a wealth of capabilities-based planning guidance in various publications, including the NRF and the National Planning Scenarios. Our findings based on review of the national policy context (as discussed in Chapters Two and Three) and interviews with military and civilian planners at five U.S. sites (as discussed in Chapters Four and Five) suggest an opportunity to fill a gap in planning by developing a tool that links a prioritized risk assessment to tailored capabilities-based planning for all communities and all hazards, in an automated way that alleviates some of the planning burden for local civilian and military planners. In this section, we describe our approach to developing the tool, beginning with design features that address needs expressed by interviewees across the five sites we visited, and then follow with our approach to developing the tool.

Addressing Perceived Needs

We will design the planning support tool to meet perceived user needs, as described earlier in this chapter.

Functional Characteristics.

Integration of Tools. Civilian and military actors are aware of the extensive inventory of existing decision support tools to support local disaster preparedness. As one interviewee commented, "Every time I go to a conference, I get a whole compact disc with a bunch of models on it." In designing the new tool, we will not replicate capabilities already available in existing tools; rather, we will design our decision support tool to leverage existing models whenever desirable.

Applicability to Planning. We will design the tool to automate linkages for planning activities across disaster phases. Specifically, we will link risk assessment with capabilities-based planning and resources needed, as shown on Figure 6.1, and we may indicate exercises that can increase preparedness. We will not specifically address budgeting or logistics with the tool. These activities are very important but take place when civilian and military planners determine resource solutions to capabilities-based planning. Resource solutions will vary widely by community based on such factors as the relative strengths and capabilities a community already has.

Scope. We will design the tool to be applicable to all U.S. communities, regardless of size. Users will input community characteristics regarding population, geography, infrastructure, and proximity to other communities. Because users can vary these inputs, the tool can be used across communities with diverse demographics.

Implementation Features.

Ease of Use. The new tool will require only a moderate amount of technical expertise, making it easy for local civilian and military planners to use. We will design the tool in a Microsoft Excel framework. We expect that most planners will be familiar with this framework and use it frequently.

Access. We intend to create a tool that can be widely distributed to communities across the United States. There should be no barrier to gaining access to the tool. We understand that local civilian and military planners who use the tool to create a plan will likely populate the tool with some local data and may desire to limit access to these data as well as to the resulting outputs. We suggest that local planners provide a computing environment that allows authorized users the necessary level of access, similar to other official or proprietary data. We will not explicitly address classified access with the tool.

Portability. We will design the tool to execute on nearly any computing platform, so it should be laptop-portable. We are designing the tool to be most useful during the planning stage of readiness planning, rather than the event management stage. For this purpose, laptop portability should be sufficient.

Security. We are designing the tool to run with Microsoft Excel wherever the software is installed. If military officials would like to use the tool in a secure environment, they can easily do so. Nevertheless, the intention behind the design is that military planners will use the tool collaboratively with civilian officials in environments that may not be appropriate for classified information. In these cases, it will be helpful if military officials can first build a plan for installation security and bring non-classified results to the collaborative environment to see how civilian and military agencies can create a cooperative plan for the whole community.

Desired Capabilities.

Automation of Guidelines. This will be a key feature of the planning support tool. We understand that local planners have difficulty following potentially useful guidance because it may be lengthy, not tailored to their communities, not from a familiar source, or potentially conflicting with other guidance. We will document our use of guidance when we populate the tool with planning factors for capabilities and resources needed. In the users' manual for the tool, we will cite our data sources so users can understand the extent to which the tool output reflects their guidance.

Integration of Source Documents. In addition to automating use of planning guidance, we will link recommendations from the tool to their sources, for local planners to understand. The tool will show the extent to which guidance allows or compels agencies to cooperate on readiness planning or prohibits them from doing so.

Updated Lists of Current Actors. We will include a capability for local planners to populate rosters of civilian and military actors within the community who have been identified as being involved with disaster preparedness planning. These data may be useful to local planners who are new to the responsibility and unfamiliar with their potential planning partners. However, it will inevitably fall to local users to populate the roster with detailed contact information because these data are likely not available nationally to pre-populate the tool.

Capabilities-Based Planning. This is a key focus of the proposed new tool. This process should be thoroughly automated, based on user input. We will discuss the details of user input in the following sections. Users will input data on community risks and community characteristics, as well as the assumptions they would like to make about the analyzed scenarios. Based on this input, the tool will automate capabilities-based planning to output recommended capabilities for each scenario. In subsequent stages, the tool will link recommended capabilities to requirements, networks, and agreements, as well as to guidance and grant fulfillment guidelines.

Resource Typing. The tool will include a module to follow after capabilities-based planning. The module will indicate what FEMA-defined resource types may fulfill the prescribed capabilities. This module will give only an indication of a possible resourcing solution; the final determination will be made by local planners based on community strengths.

Incorporation of Exercise Findings into Planning. We will not explicitly address this or the broader incorporation of lessons learned, as we believe that this is well addressed with current resources, such as the LLIS. Site visit participants requested additional capabilities in the LLIS—e.g., to search lessons by functional area and to compare similar lessons in order to define best practices. We agree that these are important capabilities, but modest modifications of the LLIS would likely be more efficient than creation of a new system.

GIS Mapping. GIS may be a very useful capability for local planners to have to support event management. GIS can show the geospatial location both of a spreading disaster and of local assets—civilian and military. We are not directly building any capabilities into the RAND tool for event management and will not include GIS. Nonetheless, local planners who use the RAND tool may wish to use GIS to map the required capabilities and resources identified by our tool.

Communications, Situational Awareness, Asset Visibility, and Marshalling Resources. These capabilities are part of event management, and we will not explicitly address them with the RAND tool because we believe that tools currently available (e.g., WebEOC) largely meet these needs.

Approach to Developing the New Tool

We propose to develop a new tool that complements existing tools and facilitates local preparedness by filling gaps described by the local stakeholders we interviewed.

Complement Existing Tools. There are already some well-developed, simple models for risk assessment in wide distribution and moderate use: Kaiser Permanente's Hazard Vulnerability Analysis assessment tool; the VA's 2008 Emergency Management Program Guidebook, JSIVA, HAZUS-MH, EMCAPS, and CATS.

Since our vision for a risk assessment tool is quite similar to some of these tools, we do not feel that development of an alternate but similar tool would provide the greatest marginal benefit. We recommend that local planners use an existing risk assessment tool then link the outputs of their risk assessment into the RAND tool to generate capabilities-based planning and resource needs and to optional resource gap analysis and network analysis.

There are also highly developed tools to support exercises, lessons learned, and event management of major local disasters, as described in the preceding section. Event management has been very attractive to commercial software and technology developers to create and market products. In our site visit interviews, we learned of many event management software products developed by commercial firms on a for-profit venture, such as WebEOC, Previstar CPS, and LiveProcess. These tools provide many of the capabilities that interviewees cited as potentially useful, such as GIS tracking of local resources, real-time communications between community partner organizations and agencies, and hazard tracking. Some interviewees appeared unaware of some of these tools already potentially available to them. Not all of these tools may be a perfect solution for local planners and emergency managers who have expressed a desire for additional capabilities, such as real-time texting. Also, many of these tools may be expensive for local communities to acquire. However, these tools are highly developed, and we do not intend to produce a real-time event management tool that will significantly improve on the capabilities provided by these currently available resources. Nor do we aim to replicate resources related to exercises or lessons learned, such as DHS's HSEEP or LLIS, respectively. Interviewees indicated that they would like a more searchable database for lessons learned, based on capabilities or functions rather than date and place of exercises and events. However, we believe that it will be better to consider improvements to the DHS LLIS than for RAND to help develop an alternative to it. We understand that a database would help local planners and suggest that this development would make a worthwhile independent research project.

Fill Gaps in Local Planning: Proposed RAND Tool. Although it may be informed by activities outside of the planning function, the proposed planning support tool will focus on enhancing local planning, as depicted in Figure 6.2 by thick black lines. Beneficiaries of the information generated by the tool will include both high-level decisionmakers and operational personnel, such as emergency managers and others.

We are designing a planning support tool to aid local planners in all communities, in all organizations. Such a tool would provide a common baseline that planners in communities across the country can use. But we hope that the tool will be used in cooperative planning sessions with representatives from military and civilian organizations, including the VA, who share a mission for community emergency management. We will design the tool in an electronic format that can include all the necessary data.

Documentation of Key Actors. As a first step to cooperative planning, planners must be acquainted with their counterparts in other civilian and military organizations. We learned in our site visit interviews that some local planners have had difficulty identifying their counterparts—specifically, identifying comparable officials in civilian and military organizations. Addressing this need, the tool will include a roster of positions that emergency planning officials may hold at local organizations. These will also be the basis for the social networking tool described in Chapter Five and later in this chapter. There will certainly be some variance across communities, which agencies are present, and how they organize their staff. But we expect that a simple guideline that indicates roles for emergency management at common civilian and military organizations can be a starting point for local planners to begin cooperative planning in communities where these networks do not exist.

Hardware and Software Requirements. As noted previously, the RAND planning support tool needs to run in almost any computing environment and require only a minimal amount of user computing experience, data entry, and time to use. The model must also be easy to distribute, entail little or no cost, have minimal hardware and software requirements,

Figure 6.2
Proposed Focus of the RAND Planning Support Tool

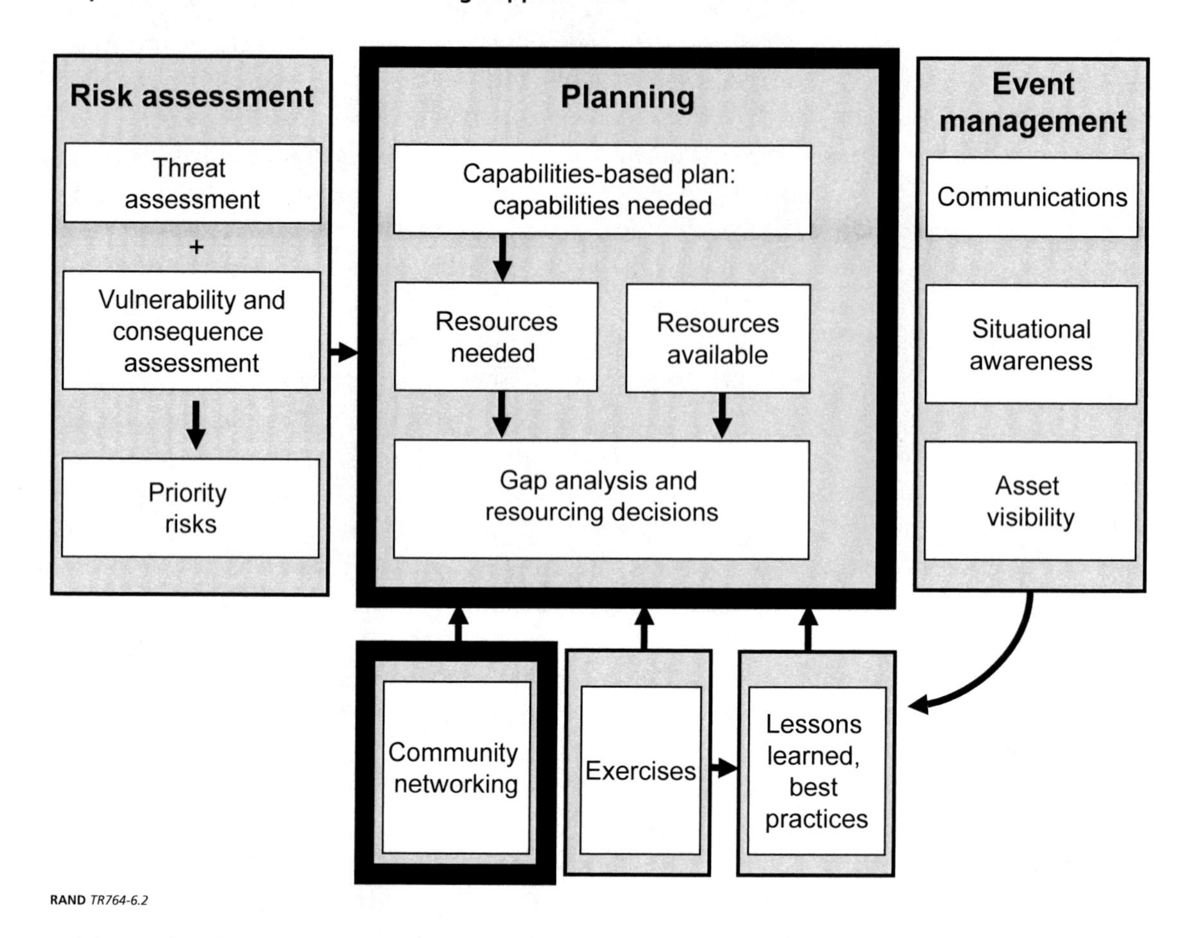

RAND *TR764-6.2*

and be usable by a wide range of individuals. To meet these design requirements, we will design the tool in a Microsoft Excel environment.

The tool can be used independently by organizations, but we also intend the tool to facilitate collaborative planning across organizations. To aid collaborative planning, users can work in a number of ways: use the tool on a single computer during a meeting, either in person or via data conference; distribute populated copies of the tool via email; or, if feasible, access a single copy of the tool via password-protected website. We will research the feasibility of this last option.

Framework for the RAND Planning Support Tool

The RAND tool will assist local planners by automating four key outputs. The tool will (1) automate the process of linking risk assessment to capabilities-based planning; (2) generate resources needed, with cost estimates when available; and, on an optional basis, (3) perform a gap analysis between resources needed and resources available and (4) generate a community disaster preparedness network map, highlighting networking opportunities and a roster of key actors. The first two of these are the core of the proposed tool and will be valuable for any set of users—either independently or across agencies. The third component—gap analysis and resourcing decision—will provide important outputs for users who are willing to enter needed input data. In our interviews, we learned that a necessary first step to community-wide collaboration is creating the network of participants. For the fourth output, we will design a tool function to give community participants visibility of their network, putting the contacts in place to facilitate collaborative use of the first three functions. We will illustrate the process of using the RAND tool, drawing on Figure 6.3 as a guide. Later in this chapter, we provide a practical example that works through the first two of these four steps.

(1) Capabilities-Based Plan: Generation of Recommended Capabilities

User Inputs: Risk Assessment. As discussed in the preceding section, planners should begin their preparedness planning by performing a risk assessment for their community, using one of the available tools. This tool should walk users through a process of identifying which scenarios are a significant threat to their communities; these scenarios are likely to occur with a relatively high probability or expected to have serious consequences for the community (or both).

Next, users should use the risk assessment tool to decide to which scenarios their communities are most vulnerable. These scenarios could be those that would pose the greatest threats to damage infrastructure or population in the community, or users may prefer to select specific scenarios of concern for planning purposes, regardless of their quantitative risk score. Users should use the tool to assess scenarios with greatest consequence (or of greatest concern, for whatever reason) for their community. These user inputs should combine to create a final score

Figure 6.3
Proposed Inputs and Outputs for the RAND Planning Support Tool

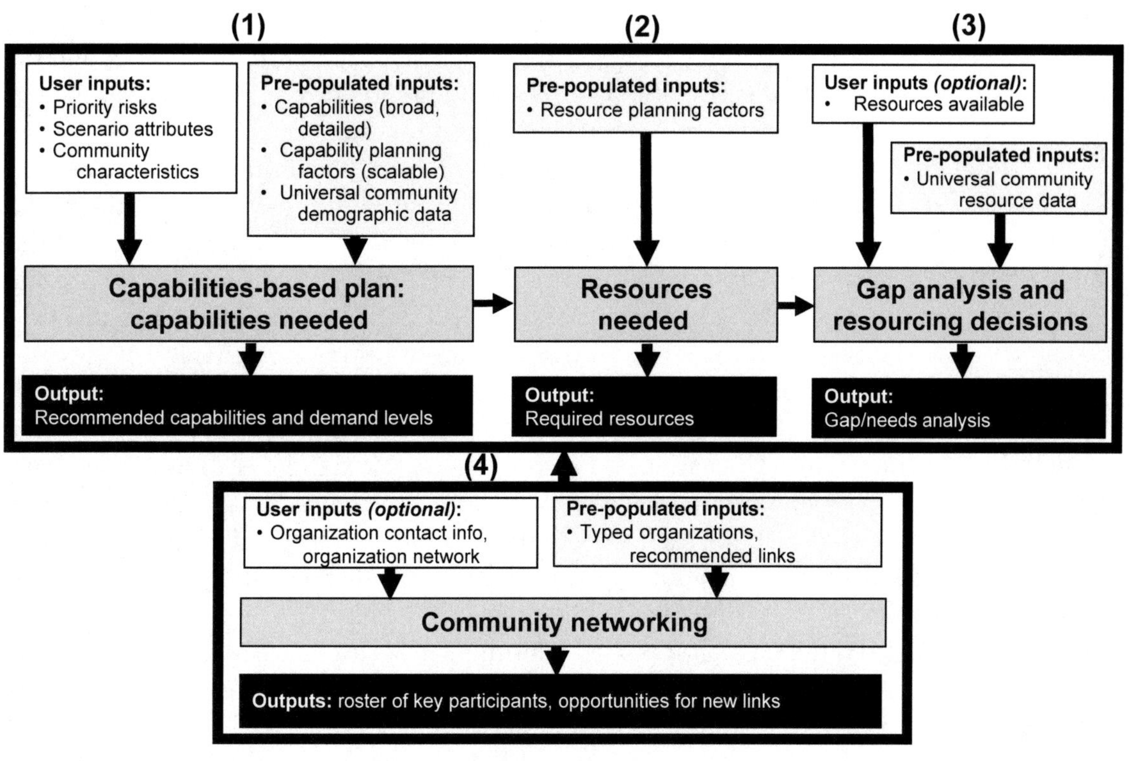

RAND *TR764-6.3*

for each scenario, determining which scenarios are of greatest total risk to a community, or at least which ones are of greatest concern to them for planning purposes. Some risk assessment tools may create output describing the consequences suffered by the community. When possible, users should note these outputs for all scenarios considered. These data may be input to the RAND tool to help guide the requirements generation process.

We are aware that some communities may already have high risk scenarios in mind based on prior information or experience, and some communities may wish to prepare for scenarios not specifically defined in existing risk assessment tools. For the next step in the planning spectrum—entering initial preferences into the RAND planning tool—local planners can easily specify whichever scenarios they determine. This can be the direct result of using an existing risk assessment tool or other community knowledge. For the second case, there will be a small burden on the users to specify unique scenarios for their community. We will describe this shortly in the following description of how users will link risk assessment to the initial stages of the RAND planning tool.

User Inputs: Community Characteristics. The tool will include pre-populated data from national-level databases, such as the Area Resource File. The tool can generate capabilities and resources needed with internally populated data. However, users may provide additional pertinent community characteristics that will allow generation of required capabilities that are scaled more finely to the needs of the local community.

Pre-Populated Inputs. To generate required capabilities based on user risk assessment inputs, we will pre-populate the RAND tool with capabilities (both broad and detailed). The tool will generate the span of capabilities required for a community to be prepared for each scenario. We will also pre-populate the tool with scaled planning factors for each of the capabilities. So, in addition to the list of required capabilities, the tool will generate the estimated amount required of each capability, based on user-input scenario characteristics and user-input community demographics.

We will use national planning guidance, such as the National Planning Scenarios and the TCL, to pre-populate the tool with these data. We will look to other sources as well, such as military guidance, and, in our research, we will generate additional data.

Output: Capabilities Needed. The first phase of the RAND tool is to link risk assessment to capabilities-based planning. Based on our analysis of our site visit interviews, we believe that this area of preparation is not supported by any existing tools. We also perceived that local planners would like planning support in this area. The National Planning Scenarios provide some support for linking capabilities to scenarios, but the RAND tool will automate this process and enhance the data, tailoring the recommendations to community characteristics.

As noted earlier, the local planner will enter the high risk scenarios into the RAND tool. The tool will allow all the scenarios in the National Planning Scenarios set as inputs. We may program some additional likely scenarios if that appears useful and if data can be verified with a particular community. However, we will also allow local planners to input custom scenarios that they deem to be high risk for their community, so long as the local planners can also supply data to support the next step of the planning process, which is linking risk assessment to plans—in particular, to capabilities-based planning. This is the link between the risk assessment activity blocks and the capabilities-based planning blocks in Figure 6.2.

The tool will add significant value to the preparedness planning process at this stage by indicating, at both a high level and a detailed level, what types of capabilities a community should have to prepare for disasters. We will populate the tool with data to match the capabilities required for a community to respond to each scenario. We will gather data from existing federal and military source documents, such as the National Planning Scenarios, the TCL, and installation preparedness guidelines. We will gather data from past event responses and exercises, in AARs and lessons learned. We will also create new data from interviews with subject matter experts, in order to populate the tool with planning factors scalable to community needs.

(2) Generation of Required Resources

Pre-Populated Inputs. The second phase of the RAND tool will generate resource requirements from capabilities requirements. We will populate the tool with data from existing sources that link capabilities to resources—human and material resources. FEMA has generated such data, as have other military, federal, and academic sources. In our research, we will define additional links as necessary.

Output: Resources Needed. We understand that the concept of capabilities-based planning explicitly allows more than one set of resources to generate a capability. So users need to decide the optimal set of resources to generate a capability in their communities. However, the tool will provide an example set of resources as an initial planning set for communities. These will be the *required resources.*

The required resources will be useful in the next phase of the tool, in which users can perform a gap analysis and produce reports identifying opportunities to increase capabilities in their communities.

(3) Gap Analysis

User Inputs: Resources Available. As a final and optional step toward planning for the capabilities and resources needed in a local area, the RAND tool will compile all the cumulative output, describing capabilities, requirements, and guidance in a format that local planners can use to execute a gap analysis. Inherently, a need analysis requires users to input data about capabilities and resources present in the community in order to calculate the difference. We understand that some users may not wish to undertake this burden, or some users may not have all the data required. So the tool will be designed with this as an optional feature.

Output: Gap Analysis. As just noted, the RAND tool will provide an opportunity for users to input data that describe current community capabilities and resources, so that the tool can perform a gap analysis. The tool will output reports to help users identify opportunities to increase capabilities in their communities. If users have not entered data reflecting capabilities and resources present in their communities, then the tool output will include the complete bottom-up need assessment. Users can process these data further however they choose.

(4) Community Network Analysis

User Inputs: Current Community Network. The RAND tool will also feature an optional module that allows local planners to map their functional emergency preparedness and response network. User inputs will be based on a network-assessment tool. The tool will include a range of functional local positions and actors in the military, VA, and other civilian agencies and will ask users (ideally, each designated position) to indicate the strength of the relationships with the other actors based on the frequency and nature of their communications and the existence or not of formal arrangements for cooperation across their agencies in disaster preparedness planning and disaster response.

Outputs: Generating Networking Opportunities and a Roster of Key Participants. Based on these inputs, we will provide existing software that will generate preprogrammed quantitative network statistics and qualitative pictures of local networks (see Chapter Five and Appendix G for more detailed descriptions and examples of networks from our five site visits). These, in turn, will indicate a roster of key community actors (and their contact information), as well as opportunities to improve local networking (e.g., by adding new actors or making some connections more direct).

Initial Vetting of the Proposed Framework

We convened several vetting meetings via teleconference to get feedback on the proposed framework for the RAND planning support tool. We invited participants from the site visits to attend the meetings. In our first rounds of vetting, we met with representatives from six organizations, spanning four of the sites. Some participants included their staff in the vetting meetings, and some teleconferences included participants from multiple locations. We were able to meet with participants from multiple locations, civilian participants, and participants from military installations across multiple military services. We received widely varying feedback across the teleconferences. We next describe the feedback and attribute it to types of site visit participants.

Capabilities-Based Planning/Requirements Generation

Some site visit participants expressed interest in using the capabilities-based planning functions. These site visit participants were from civilian organizations, and we can presume that they did not have processes in place to automate capabilities-based planning, nor detailed guidance describing which capabilities their organizations needed to support.

Two site visit participants said that they were not interested in using the capabilities-based planning functions. Some of these participants worked in civilian organizations and were very experienced working in their position. Presumably, these participants felt confident that they understood the capabilities that their organizations should provide to sustain community preparedness. Other participants who were not interested in capabilities-based planning functions worked at military installations and received very clear guidance from higher-level commands, describing the capabilities they needed to support their installations for preparedness.

Burden of Use

Site visit participants who were interested in using the capabilities-based planning functions stated that it seemed like the tool would be reasonable to use and that the outputs would be at an appropriate level of detail. Site visit participants who were not interested in the capabilities-based planning functions seemed concerned that the activity of convening representatives from across community organizations would simply be an additional burden in their work schedule. In our future vetting sessions, we will present ways in which community officials can use the tool via data conferencing and remote collaboration. We will also discuss how the tool can be useful even to officials for planning at their individual organizations.

Cooperative Planning Environment

We learned from the participants that the extent to which communities meet to plan cooperatively varies widely. Organizations in some communities do not meet frequently as a group to discuss planning strategies. In such communities, there may be many jurisdictions, and the jurisdictions may not have a structured process for collaborating on preparedness planning.

In other communities, we learned that the government and emergency management organizations representing the larger geographic community—for instance, the county—are very well linked and work aggressively to include military organizations as well as other civilian organizations in the community. In all cases, the site visit participants expressed interest in increasing coordination across organizations in the community, military and civilian. They broadly expressed interest in understanding what capabilities each organization can support during an emergency event. They also expressed interest in participating more frequently in other organizations' exercises. The participants expressed dismay that they were not more closely acquainted with their counterparts at other organizations and that they were poorly informed of upcoming exercises.

Community Network Assessment

We found strong interest in a tool function to strengthen community networks, particularly by helping officials contact their counterparts at other organizations and by increasing community awareness and collaborative participation in exercises. We are encouraged that a community network function, keeping a list of organizations and officials contact information, could be a valuable asset to communities. We propose that the same Internet-based portal could include a calendar function that community officials can populate in order to increase awareness of local training and exercises.

Gap-Analysis Function

We did not receive significant feedback on the gap-analysis function. Participants were more interested in providing feedback on the prior tool functions and did not explicitly discuss whether they would be likely to input data from their organization to measure gaps in capabilities. We expect that, after the other tool functions are more fully developed, we can conduct further vetting to see how likely users would be to utilize the gap-analysis function.

Other Desired Functions

In the vetting teleconferences, participants reiterated several functional requests that arose during the site visits. Participants stated that they hoped that a RAND-developed tool would address needs in their communities to improve communications between organizations. They cited examples in which organizations in their communities provided very similar capabilities, such as fire protection or emergency medical response, but the organizations had separate communications systems and could not easily coordinate a response to an event.

Participants also wanted a tool that gave organizations visibility into what capabilities or resources other organizations could supply in response to an event. This desire would be particularly acute when one organization would be dependent on another for support in that capabilities area.

Vetting Conclusions

Reactions to the capabilities-based planning tool functions were mixed. We believe that some site visit participants will become more interested in the capabilities-based planning function once they see a more fully developed proof of concept. We also found that organization officials without any support for capabilities-based planning, or with limited experience in the emergency manager position, were supportive of the capabilities-based planning function. In contrast, participants who were not enthusiastic were quite experienced in their position or received clear capabilities-based planning guidance from higher-level military commands.

Our site visits were conducted in communities with strong military presences, most of which had also received substantial grant funding in the civilian sector for emergency preparedness. We believe that, in a broader sample of communities across the United States, we might find a larger proportion of military installations that are less advanced in their capabilities-based planning activities and of civilian officials who are not as expert emergency managers as the senior managers we interviewed during this formative research phase. In these instances, we would expect to see a greater interest in a tool to facilitate capabilities-based planning. We plan to field test the prototype tool in a broader range of communities, including smaller communities with and without major military installations.

Example of the Proposed Tool: Generation of Capabilities Needed and Resources Needed

In this section, we present an example of how the proposed planning support tool to be developed by RAND researchers under this project would work for the first two of the proposed four tool outputs: capabilities needed and resources needed. For this example, we show the data that a user should enter into the RAND planning tool—in addition to pre-populated tool data—and potential tool outputs. The pre-populated data in this example are extracted from the TCL.

As discussed earlier in this chapter, the initial risk assessment data required to perform capabilities-based planning must be input by the user. In Figure 6.4, we show data inputs related to this step; both of the first two columns need to be input by users. Users should input

Figure 6.4
Initial Data Elements for the RAND Planning Tool

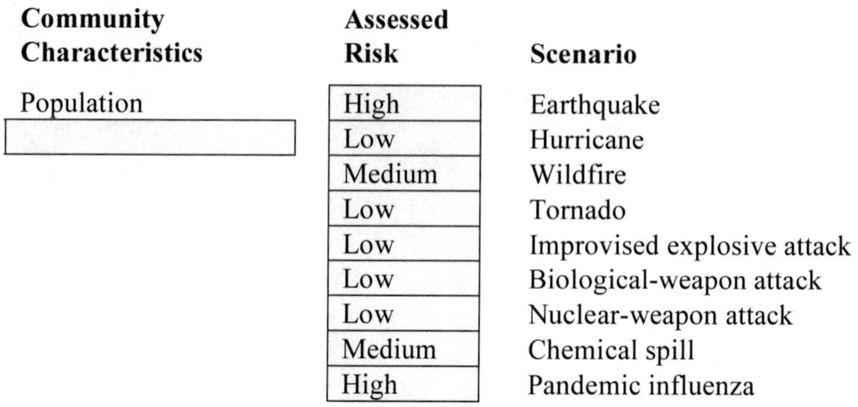

Community Characteristics	Assessed Risk	Scenario
Population	High	Earthquake
	Low	Hurricane
	Medium	Wildfire
	Low	Tornado
	Low	Improvised explosive attack
	Low	Biological-weapon attack
	Low	Nuclear-weapon attack
	Medium	Chemical spill
	High	Pandemic influenza

RAND *TR764-6.4*

characteristics about their communities in order to scale the needed capabilities. In this example, we ask users to input the population of their communities. In the tool, we will include some pre-populated data and will also ask for additional community characteristics that will help users tailor the planning more precisely to their local circumstances (e.g., plan for a geographic area not captured explicitly in population data) and characteristics that are not available through public databases.

The second piece of data that users must input is a risk assessment of hazards in their communities. As shown in Figure 6.5, the RAND tool will be pre-populated with a list of scenarios that communities should consider. Users should input their risk assessment of the scenarios considered in the tool. We will provide users a range of risk assessment options from which to select; a scale of high to low risk may be a good scale to use. We will include additional scenarios in the tool; Figure 6.5 includes only a few examples. The RAND tool will also allow users to include custom or hybrid scenarios specific to their community in the capabilities-based planning activity. We do not portray that feature in this example. We will also ask users to input attributes for each of the scenarios but do not explicitly include such an example here.

The next activity in the capabilities-based planning stage of the RAND tool links capabilities to scenarios; the RAND tool will automatically generate the list of capabilities needed for each scenario considered. The user-input data for risk assessment shows that the pandemic influenza scenario is assessed to have high risk. In Figure 6.5, we show examples of capabilities that we will pre-populate in the tool. For each scenario selected by the user, the tool will generate a set of capabilities needed. For the pandemic influenza scenario, we will show an example of capabilities needed for medical surge.

In Figure 6.6, we see how the tool will use pre-populated planning factors to estimate the amount of each capability needed, based on community characteristics and scenario attributes. The capabilities needed are quantified in terms of units specific to the capability. In this example, we consider the capabilities needed for medical surge. Medical surge capabilities can be measured in terms of the number of patients, so the scaled capabilities needed in this

Figure 6.5
Pre-Populated Tool Data Describing Scenarios and Capabilities

Scenarios

- Earthquake
- Hurricane
- Wildfire
- Tornado
- Improvised explosive attack
- Biological-weapon attack
- Nuclear-weapon attack
- Chemical spill
- Pandemic influenza

Capabilities

- Animal-disease emergency support
- Environmental health
- Explosive-device response operations
- Fire-incident response support
- WMD and hazardous materials
- Response and decontamination
- Citizen evacuation and shelter in place
- Isolation and quarantine
- Search and rescue (land based)
- Emergency public information and warning
- Emergency triage and prehospital treatment
- Medical surge
- Medical-supply management and distribution
- Mass prophylaxis
- Mass care (sheltering, feeding, and related services)

NOTE: WMD = weapons of mass destruction.

RAND *TR764-6.5*

Figure 6.6
Example of the Amount of Capabilities Needed for Medical Surge

Capabilities	Capabilities needed
Animal-disease emergency support Environmental health Explosive-device response operations Fire-incident response support WMD and hazardous materials Response and decontamination Citizen evacuation and shelter in place Isolation and quarantine Search and rescue (land based) Emergency public information and warning Emergency triage and prehospital treatment **Medical surge** Medical-supply management and distribution Mass prophylaxis Mass care (sheltering, feeding, and related services)	• 73,300 patients hospitalized – 20% (14,600) are critical and require a care bed, mechanical ventilation – 80% (58,640) are noncritical – 1% require transport to specialty clinic >100 miles distant • Outpatient visits • Home self-care patients

RAND *TR764-6.6*

example are listed in terms of patients. We provide in Figure 6.6 a partial list of the detailed capabilities needed in the broad capabilities area of medical surge. We display this tool output on the right. These data represent the potential final output for the capabilities-based planning stage of the tool.

In Figure 6.7, we show the automated tool function that links capabilities to resources. We show the data elements underlying the tool function generating resources needed.

Recall that a community may use any combination of resources to support a needed capability; the tool will output one possible solution. We see in Figure 6.7 that resources needed can be defined in terms of personnel, equipment, and supplies, as well as sets or teams of these resources. We see that, in order to support the capabilities needed in this example, a community could supply a quantity of registered nurses (RNs), medical doctors (MDs), and respiratory therapists, as well as specific prescription and over-the-counter (OTC) pharmaceuticals. If

Figure 6.7
Output Data: Capabilities Needed and Resources Needed

Capabilities needed	Resources needed
• 73,300 patients hospitalized	
– 20% (14,600) are critical and require a care bed, mechanical ventilation	7,330 RNs 1,466 MDs 2,932 respiratory therapists
– 80% (58,640) are noncritical	2,932 RNs 1,466 MDs
– 1% require transport to specialty clinic >100 miles distant	Variable depending on patient condition
• Outpatient visits	Antivirals, antipyretics, analgesics
• Home self-care patients	OTC drugs

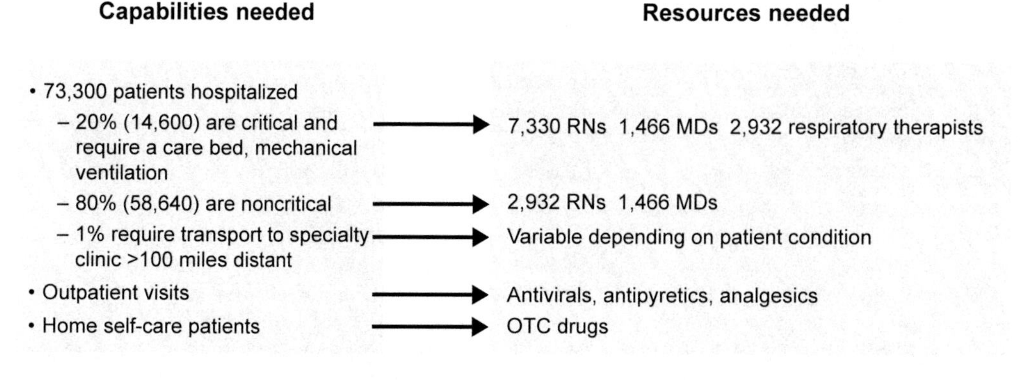

NOTE: RN = registered nurse. MD = medical doctor. OTC = over the counter.
RAND *TR764-6.7*

users opt to input data describing the capabilities and resources present in their communities, the RAND tool can compare these data inputs to the tool-generated outputs on capabilities needed and resources needed and perform the next tool function, a gap analysis. (That step and the community networking step are not depicted here.)

CHAPTER SEVEN

Summary and Next Steps

In this report, we have described our review of current policies and programs—especially those applicable nationwide—under which local disaster preparedness now operates. We have also discussed the importance of sound concepts and processes for conducting effective risk assessments and capabilities-based planning.

We have described site visits at five locations to learn how communities actually prepare for disasters and to identify, through extensive interviews with civilian and military entities, the desired features and capabilities for a new planning support tool.

We then explained how we have integrated our understanding of both the policy context and of local preparedness planning needs to develop a framework for a local capabilities-based planning support tool and complementary efforts to assess and improve connections among relevant local agencies. Using data from the interviews and our research into existing preparedness support methodologies, we identified a clear gap that the proposed RAND planning support tool can fill. We propose to develop a prototype tool that will assist with community planning for disasters, focusing on risk-informed and capabilities-based planning. We also propose to develop a tool function that will enable local planners to strengthen their community networks. The tool will create an environment for community organizations to share contact information, help users identify key organizations with which they can partner to coordinate capabilities should an event occur, and provide an environment in which organizations can share information about upcoming exercises.

The next steps in our research effort will be to develop and field test a prototype tool. It will be a workable tool—capable of testing several inputs and outputs. Based on the level of time and project resources currently allocated for the next phase, this first prototype will be a usable tool to support planning for a subset of capabilities (e.g., medical capabilities) related to some but not all disaster scenarios. More time and effort will be required to develop a full-scale, fully functional tool that incorporates all capabilities across all scenarios and is ready for production and distribution. Throughout development, we will work with subject matter experts to inform the tool features, and we will continue to vet the tool with community representatives. After developing the prototype tool, we will perform a proof of concept through field tests across as broad as possible a range of U.S. communities. The proof-of-concept field tests will help in identifying areas for improvement in the further development of the tool and thus will inform the development of the all-capabilities all-hazards planning support tool that is the ultimate goal of this effort.

APPENDIX A

Definitions and Terminology

To avoid ambiguity and to ensure consistency in discussion, the following are definitions of terms used in this report. When a definition is from an official publication, we provide a citation to it.

all-hazards. Describes an incident, natural or manmade, that warrants action to protect life, property, environment, and public health or safety and to minimize disruptions of government, social, or economic activities (derived from NRF).

assessment. The evaluation and interpretation of measurements and other information to provide a basis for decisionmaking (derived from NRF).

capability. Provides the means to accomplish a mission or function. A capability results from the performance of one or more critical tasks, under specified conditions, to target levels of performance. In joint military doctrine, *capability* means the ability to execute a specified course of action. A capability may be delivered with any combination of properly planned, organized, equipped, trained, and exercised personnel that achieves the desired outcome.

capabilities-based preparedness, capabilities-based planning. Preparing, under uncertainty, to provide capabilities suitable for a wide range of challenges. It recognizes the finite nature of resources within an economic framework that necessitates prioritization and choice of how a decisionmaker will execute a mission and employ capabilities within his or her control.

catastrophic incident. Any natural or manmade incident, including terrorism, that results in extraordinary levels of mass casualties, damage, or disruption severely affecting the population, infrastructure, environment, economy, national morale, and/or government functions (derived from NRF).

Citizen Corps. A community-level program, administered by the Department of Homeland Security, that brings government and private-sector groups together and coordinates the emergency preparedness and response activities of community members. Through its network of community, state, and tribal councils, Citizen Corps increases community preparedness and response capabilities through public education, outreach, training, and volunteer service (derived from NRF).

consequences. Damages, injuries, and fatalities that result from disaster or other emergency incidents. Such incidents may be either natural or manmade, including both intentional acts and accidents.

coordinate. To systematically advance an analysis and exchange of information among principals who have or may have a need to know certain information to carry out specific incident management responsibilities (derived from NRF).

critical infrastructure. Systems, assets, and networks, whether physical or virtual, so vital to the United States that the incapacity or destruction of such systems and assets would have a debilitating impact on security, national economic security, national public health or safety, or any combination of those matters (derived from NRF).

disaster. Any incident, natural or manmade, that causes significant loss of life or personal injury, severe population disruptions, major interference with the orderly functioning of government, or significant damage to property, the economy, or the environment. A disaster in this context may or may not rise to the level of "major disaster" as defined in the Stafford Act.

emergency. Any incident, whether natural or manmade, that requires responsive action to protect life or property. Under the Robert T. Stafford Disaster Relief and Emergency Assistance Act, an emergency means any occasion or instance for which, in the determination of the President, federal assistance is needed to supplement state and local efforts and capabilities to save lives and to protect property and public health and safety, or to lessen or avert the threat of a catastrophe in any part of the United States (derived from NRF).

emergency management. As subset of incident management, the coordination and integration of all activities necessary to build, sustain, and improve the capability to prepare for, protect against, respond to, recover from, or mitigate against threatened or actual natural disasters, acts of terrorism, or other manmade disasters (derived from NRF).

Emergency Management Assistance Compact (EMAC). A congressionally ratified organization that provides form and structure to interstate mutual aid. Through EMAC, a disaster affected state can request and receive assistance from other member states quickly and efficiently, resolving two key issues up front: liability and reimbursement (derived from NRF).

emergency manager. The person who has the day-to-day responsibility for emergency management programs and activities. The role is one of coordinating all aspects of a jurisdiction's mitigation, preparedness, response, and recovery capabilities (derived from NRF).

emergency operations center (EOC). The physical location at which the coordination of information and resources to support incident management (on-scene operations) activities normally takes place. An EOC may be a temporary facility or may be located in a more central or permanently established facility, perhaps at a higher level of organization within a jurisdiction. EOCs may be organized by major functional disciplines (e.g., fire, law enforcement, medical services), by jurisdiction (e.g., federal, state, regional, tribal, city, county), or some combination thereof (derived from NRF).

emergency plan. The ongoing plan maintained by various jurisdictional levels for responding to a wide variety of potential hazards (derived from NRF).

evacuation. Organized, phased, and supervised withdrawal, dispersal, or removal of civilians from dangerous or potentially dangerous areas, and their reception and care in safe areas (derived from NRF).

federal agency. "Any department, independent establishment, government corporation, or other agency of the executive branch of the federal government, including the United States Postal Service, but shall not include the American National Red Cross" (Stafford, 1988).

Fusion Center. Facility that brings together into one central location law enforcement, intelligence, emergency management, public health, and other agencies, as well as private-sector and nongovernmental organizations when appropriate, and that has the capabilities to evaluate and act appropriately on all available information (derived from NRF).

hazard. Something that is potentially dangerous or harmful, often the root cause of an unwanted outcome (derived from NRF).

Homeland Security Exercise and Evaluation Program (HSEEP). A capabilities- and performance-based exercise program that provides a standardized methodology and terminology for exercise design, development, conduct, evaluation, and improvement planning (derived from NRF).

Homeland Security Information Network (HSIN). The primary reporting method (common national network) for the Department of Homeland Security to reach departments, agencies, and operations centers at the federal, state, and local levels and in the private sector. HSIN is a collection of systems and communities of interest designed to facilitate information sharing, collaboration, and warnings (derived from NRF).

incident. An occurrence or event, natural or manmade, that requires a response to protect life or property. Incidents can, for example, include major disasters, emergencies, terrorist attacks, terrorist threats, civil unrest, wildland and urban fires, floods, HAZMAT spills, nuclear accidents, aircraft accidents, earthquakes, hurricanes, tornadoes, tropical storms, tsunamis, war-related disasters, public health and medical emergencies, and other occurrences requiring an emergency response (derived from NRF).

Incident Command System (ICS). A standardized on-scene emergency management construct specifically designed to provide for the adoption of an integrated organizational structure that reflects the complexity and demands of single or multiple incidents, without being hindered by jurisdictional boundaries. ICS is a management system designed to enable effective incident management by integrating a combination of facilities, equipment, personnel, procedures, and communications operating within a common organizational structure, designed to aid in the management of resources during incidents. It is used for all kinds of emergencies and is applicable to small as well as large and complex incidents. ICS is used by various jurisdictions and functional agencies, both public and private, to organize field-level incident management operations (derived from NRF).

interoperability. The ability of emergency management/response personnel to interact and work well together. In the context of technology, *interoperability* also refers to having an emergency communications system that is the same or is linked to the same system that a jurisdiction uses for nonemergency procedures and that effectively interfaces with national standards as they are developed. The system should allow the sharing of data with other jurisdictions and levels of government during planning and deployment (derived from NRF).

Joint Field Office (JFO). The primary federal incident management field structure. The JFO is a temporary federal facility that provides a central location for the coordination of federal, state, tribal, and local governments and private-sector and nongovernmental organizations with primary responsibility for response and recovery. The JFO structure is organized, staffed, and managed in a manner consistent with National Incident Management System principles and is led by the Unified Coordination Group. Although the JFO uses an Incident Command System structure, the JFO does not manage on-scene operations. Instead, the JFO focuses on providing support to on-scene efforts and conducting broader support operations that may extend beyond the incident site (derived from NRF).

jurisdiction. A range or sphere of authority. Public agencies have jurisdiction at an incident related to their legal responsibilities and authority. Jurisdictional authority at an incident can be political or geographical (e.g., federal, state, tribal, local boundary lines) or functional (e.g., law enforcement, public health) (derived from NRF).

local government. "A county, municipality, city, town, township, local public authority, school district, special district, intrastate district, council of governments (regardless of whether the council of governments is incorporated as a nonprofit corporation under state law), regional or interstate government entity, or agency or instrumentality of a local government; an

Indian tribe or authorized tribal entity, or in Alaska a Native Village or Alaska Regional Native Corporation; a rural community, unincorporated town or village, or other public entity" (Pub. L. 107-296).

manmade, man-made. As distinguished from natural disasters, incidents that are caused by people: criminal activity (including terrorism); other noncriminal intentional activity; and the full range of incidents that occur by accident (including industrial, transportation, environmental, and recreational).

major disaster. Under the Robert T. Stafford Disaster Relief and Emergency Assistance Act, any natural catastrophe (including any hurricane, tornado, storm, high water, wind-driven water, tidal wave, tsunami, earthquake, volcanic eruption, landslide, mudslide, snowstorm, or drought) or, regardless of cause, any fire, flood, or explosion in any part of the United States that, in the determination of the President, causes damage of sufficient severity and magnitude to warrant major disaster assistance under the Stafford Act to supplement the efforts and available resources of states, local governments, and disaster relief organizations in alleviating the damage, loss, hardship, or suffering caused thereby (derived from NRF).

mass casualty event. Distinct from *major disaster*, defined as any event that overwhelms the medical response and treatment capabilities of a medical facility, medical system, or community. Joint military doctrine specifically defines a mass casualty event as "any large number of casualties produced in a relatively short period of time, usually as the result of a single incident such as a military aircraft accident, hurricane, flood, earthquake, or armed attack that exceeds local logistic support capabilities" (JP 1-02).

mitigation. Activities providing a critical foundation in the effort to reduce the loss of life and property from natural and/or manmade disasters by avoiding or lessening the impact of a disaster and providing value to the public by creating safer communities. Mitigation seeks to fix the cycle of disaster damage, reconstruction, and repeated damage. These activities or actions, in most cases, will have a long-term sustained effect (derived from NRF).

mutual aid or assistance agreement. Written or oral agreement between and among agencies, organizations, or jurisdictions (or any combination thereof) that provides a mechanism to quickly obtain emergency assistance in the form of personnel, equipment, materials, and other associated services. The primary objective is to facilitate rapid, short-term deployment of emergency support prior to, during, and after an incident (derived from NRF).

national. Of a nationwide character, including the federal, state, tribal, and local aspects of governance and policy (derived from NRF).

National Defense Area (NDA). "An area established on non-federal lands located within the United States or its possessions or territories for the purpose of safeguarding classified defense information or protecting Department of Defense (DOD) equipment and/or materiel. Establishment of a national defense area temporarily places such non-federal lands under the effective control of the Department of Defense and results only from an emergency event. The senior DOD representative at the scene will define the boundary, mark it with a physical barrier, and post warning signs. The landowner's consent and cooperation will be obtained whenever possible; however, military necessity will dictate the final decision regarding location, shape, and size of the national defense area." (JP 1-02).

National Disaster Medical System (NDMS). A federally coordinated system that augments the nation's medical response capability. The overall purpose of the NDMS is to establish a single, integrated national medical response capability for assisting state and local authorities in dealing with the medical impacts of major peacetime disasters. NDMS, under Emergency

Support Function #8—Public Health and Medical Services, supports federal agencies in the management and coordination of the federal medical response to major emergencies and federally declared disasters (derived from NRF).

National Exercise Program. A Department of Homeland Security–coordinated exercise program based on the National Planning Scenarios, which are contained in the National Preparedness Guidelines. This program coordinates and, where appropriate, integrates a five-year homeland security exercise schedule across federal agencies and incorporates exercises at the state and local levels (derived from NRF).

National Incident Management System (NIMS). System that provides a proactive approach guiding government agencies at all levels, the private sector, and nongovernmental organizations to work seamlessly to prepare for, prevent, respond to, recover from, and mitigate the effects of incidents, regardless of cause, size, location, or complexity, in order to reduce the loss of life or property and harm to the environment (derived from NRF).

National Planning Scenarios. A component of the National Preparedness Guidelines, the National Planning Scenarios depict a diverse set of high-consequence threat scenarios of both potential terrorist attacks and naturally occurring events such as natural disasters and large-scale disease outbreaks. Collectively, the 15 scenarios are designed to focus contingency planning for homeland security preparedness work at all levels of government and with the private sector. The scenarios form the basis for coordinated federal planning, training, exercises, and grant investments needed to prepare for emergencies of all types (derived from NRF).

National Preparedness Guidelines (NPG). Guidance that establishes a vision for national preparedness and provides a systematic approach for prioritizing preparedness efforts across the nation. These guidelines focus policy, planning, and investments at all levels of government and the private sector. The guidelines replace the Interim National Preparedness Goal and integrate recent lessons learned (derived from NRF).

National Response Framework (NRF). Guides how the nation conducts all-hazards response. The framework documents the key response principles, roles, and structures that organize national response. It describes how communities, states, the federal government, and private-sector and nongovernmental partners apply these principles for a coordinated, effective national response. And it describes special circumstances in which the federal government exercises a larger role, including incidents in which federal interests are involved and catastrophic incidents in which a state would require significant support. It allows first responders, decisionmakers, and supporting entities to provide a unified national response (derived from NRF).

nongovernmental organization (NGO). An entity with an association that is based on interests of its members, individuals, or institutions. It is not created by a government, but it may work cooperatively with government. Such organizations serve a public purpose, not a private benefit. Examples of NGOs include faith-based charity organizations and the American Red Cross. NGOs, including voluntary and faith-based groups, provide relief services to sustain life, reduce physical and emotional distress, and promote the recovery of disaster victims. Often, these groups provide specialized services that help individuals with disabilities. NGOs and voluntary organizations play a major role in assisting emergency managers before, during, and after an emergency (derived from NRF).

planned event: A planned, nonemergency activity (e.g., sporting event, concert, parade) (derived from NRF).

preparedness. Actions that involve a combination of planning, resources, training, exercising, and organizing to build, sustain, and improve operational capabilities. Preparedness is the process of identifying the personnel, training, and equipment needed for a wide range of potential incidents, and developing jurisdiction-specific plans for delivering capabilities when needed for an incident (derived from NRF). The range of deliberate, critical tasks and activities necessary to build, sustain, and improve the operational capability to prevent, protect against, respond to, and recover from domestic incidents. Preparedness is a continuous process. Preparedness involves efforts at all levels of government and coordination among government, private-sector, and nongovernmental organizations to identify threats, determine vulnerabilities, and identify required resources. Within the NIMS, preparedness is operationally focused on establishing guidelines, protocols, and standards for planning, training and exercises, personnel qualification and certification, equipment certification, and publication management (derived from NPG).

prevention. Actions to avoid an incident or to intervene to stop an incident from occurring. Prevention involves actions to protect lives and property. It involves applying intelligence and other information to a range of activities that may include such countermeasures as deterrence operations; heightened inspections; improved surveillance and security operations; investigations to determine the full nature and source of the threat; public health and agricultural surveillance and testing processes; immunizations, isolation, or quarantine; and, as appropriate, specific law enforcement operations aimed at deterring, preempting, interdicting, or disrupting illegal activity and apprehending potential perpetrators and bringing them to justice (derived from NRF).

Principal Federal Official (PFO). May be appointed to serve as the Secretary of Homeland Security's primary representative to ensure consistency of federal support as well as the overall effectiveness of the federal incident management for catastrophic or unusually complex incidents that require extraordinary coordination (derived from NRF).

private sector. Organizations and entities that are not part of any governmental structure. The private sector includes for-profit and not-for-profit organizations, formal and informal structures, commerce, and industry (derived from NRF).

protection. Actions to reduce the vulnerability of critical infrastructure or key resources in order to deter, mitigate, or neutralize terrorist attacks, major disasters, and other emergencies. It requires coordinated action on the part of federal, state, and local governments, the private sector, and concerned citizens across the country. Protection also includes continuity of government and operations planning; awareness elevation and understanding of threats and vulnerabilities to their critical facilities, systems, and functions; identification and promotion of effective sector-specific protection practices and methodologies; and expansion of voluntary security-related information sharing among private entities within the sector as well as between government and private entities (derived from NPG).

recovery. The development, coordination, and execution of service- and site-restoration plans; the reconstitution of government operations and services; individual, private-sector, nongovernmental, and public assistance programs to provide housing and to promote restoration; long-term care and treatment of affected persons; additional measures for social, political, environmental, and economic restoration; evaluation of the incident to identify lessons learned; post-incident reporting; and development of initiatives to mitigate the effects of future incidents (derived from NRF).

resources. Personnel and major items of equipment, supplies, and facilities available or potentially available for assignment to incident operations and for which status is maintained. Under NIMS, resources are described by kind and type and may be used in operational support or supervisory capacities at an incident or at an EOC.

response. Immediate actions to save lives, protect property and the environment, and meet basic human needs. Response also includes the execution of emergency plans and actions to support short-term recovery (derived from NRF).

risk. Risk is a function of three variables: threat, vulnerability, and consequence (derived from NPG).

risk assessment. The comprehensive process for the identification and characterization of threat, consequences, and vulnerabilities. While each element is important for capabilities-based planning and national preparedness, determinations of vulnerability are important because they include an assessment of exposure, sensitivity, and resilience. Resilience is a critical factor because it refers to an organization's or community's coping capacity to absorb events and adapt, respond to, and recover from the event's effects.

situational awareness. The ability to identify, process, and comprehend the critical elements of information about an incident (derived from NRF).

special-needs populations. Populations whose members may have additional needs before, during, and after an incident in functional areas, including but not limited to maintaining independence, communications, transportation, supervision, and medical care. Individuals in need of additional response assistance may include those who have disabilities; who live in institutionalized settings; who are elderly; who are children; who are from diverse cultures; who have limited English proficiency or are non–English speaking; or who are transportation disadvantaged (derived from NRF).

Stafford Act. The Robert T. Stafford Disaster Relief and Emergency Assistance Act, Pub. L. 93-288, as amended. This act describes the programs and processes by which the federal government provides disaster and emergency assistance to state and local governments, tribal nations, eligible private nonprofit organizations, and individuals affected by a declared major disaster or emergency. The Stafford Act covers all hazards, including natural disasters and terrorist events (derived from NRF).

state. When referring to a governmental entity, it means "any State of the United States, the District of Columbia, the Commonwealth of Puerto Rico, the U.S. Virgin Islands, Guam, American Samoa, the Commonwealth of the Northern Mariana Islands, and any possession of the United States" (Pub. L. 107-296, §2[14]).

Strategic National Stockpile (SNS). A national repository of antibiotics, chemical antidotes, antitoxins, life-support medications, intravenous administration, airway-maintenance supplies, and medical and surgical items. The SNS is designed to supplement and resupply state and local public health agencies in the event of a national emergency anywhere and at anytime within the United States or its territories.

Target Capabilities List (TCL). Defines specific capabilities that all levels of government should possess in order to respond effectively to incidents (derived from NRF).

territories. Under the Stafford Act, U.S. territories are may receive federally coordinated response within the U.S. possessions, including the insular areas, and within the Federated States of Micronesia (FSM) and the Republic of the Marshall Islands (RMI). Stafford Act assistance is available to Puerto Rico, the U.S. Virgin Islands, Guam, American Samoa, and the Commonwealth of the Northern Mariana Islands, which are included in the definition of *state* in the Stafford Act. At present, Stafford Act assistance also is available to the FSM and the RMI under the compact of free association (derived from NRF).

terrorism. As defined under the Homeland Security Act of 2002 (Pub. L. 107-296 §2[15]), any activity that involves an act dangerous to human life or potentially destructive of critical infrastructure or key resources; is a violation of the criminal laws of the United States or of any state or other subdivision of the United States in which it occurs; and is intended to intimidate or coerce the civilian population or influence or affect the conduct of a government by mass destruction, assassination, or kidnapping.

threat. An indication of possible violence, harm, or danger (derived from NRF).

tribal. Referring to any Indian tribe, band, nation, or other organized group or community, including any Alaskan Native Village as defined in or established pursuant to the Alaskan Native Claims Settlement Act (85 Stat. 688, 43 U.S.C. 1601 et seq.), that is recognized as eligible for the special programs and services provided by the United States to Indians because of their status as Indians (derived from NRF).

threat assessment. A process used to evaluate the probability of terrorist activity, natural disaster, or other type of incident on a given population, asset, or location that has the potential to exploit a vulnerability.

Universal Task List (UTL). A menu of unique tasks that link strategies to prevention, protection, response, and recovery tasks for the major events represented by the National Planning Scenarios. It provides a common vocabulary of critical tasks that support development of essential capabilities among organizations at all levels. The list was used to assist in creating the Target Capabilities List (derived from NRF).

vulnerability. A state inherent in the manifestation of physical, organizational, and cultural properties of a system that can result in damage if attacked by an adversary or subjected to a natural disaster or some other form of threat.

vulnerability assessment. A process to identify physical, organizational, or cultural characteristics or procedures that render populations, assets, areas, or special events susceptible to a specific hazard or set of hazards.

APPENDIX B

Local Disaster Preparedness in the Civilian Sector

This appendix provides greater detail about the different federal programs that underpin local civilian preparedness planning and the state-level mechanisms mentioned in Chapter Two.

Federal Structures and Programs for States

U.S. Department of Homeland Security

In addition to the national guidance contained in the NRF and the NPG, DHS has a number of assistance and grant programs designed to assist localities.

Fusion Centers. DHS has encouraged states and localities to establish Fusion Centers to share information and intelligence within the jurisdictional responsibility of the state or local entity and with the federal government. As of March 2009, there are 58 Fusion Centers around the country. DHS supports the centers through direct financial contributions (more than $259 million in FYs 2004–2007) as well as through deployment of personnel with operational and intelligence expertise. Other federal agencies (FBI, ATF, DEA) have also provided personnel to selected centers. Almost half of the Fusion Centers have access to the Homeland Security Data Network, through which states may obtain classified national security intelligence, including terrorism-related classified intelligence from the National Counterterrorism Center, an entity in the Office of the Director of National Intelligence.

Homeland Security Grant Program (HSGP). The HSGP provides funds for planning, organization, equipment, training, and exercise activities in support of the NPG and related plans and programs. HSGP includes the state Homeland Security Program, which supports building and sustaining capabilities at the state and local levels. In addition, it includes funding for the UASI, the Citizen Corps program via the Community Emergency Response Teams (CERTs), and the MMRS (all described in this section).

HSEEP. The HSEEP provides federal policy and program guidance that is intended to create a national standard for exercises. To that end, the HSEEP provides baseline terminology that DHS expects will be used by all planners (federal, state, and local) for any exercise that may involve DHS responsibilities for incident management at the federal level, and for DHS-funded and -supported exercise activities for states and localities. The HSEEP also provides tools to help exercise managers plan, conduct, and evaluate exercises, as well as lessons learned and best practices from existing and prior exercise programs. The HSEEP integrates the principles and policies contained in the NPG and the NRF, especially the National Planning Scenarios, UTL, TCL, and NIMS. One of the key tenets of the HSEEP is that it is intended to be capabilities based.

UASI. The UASI program provides additional funding to designated jurisdictions to address the multidisciplinary planning, operations, equipment, training, and exercise needs of high-threat, high-density urban areas. UASI is intended to help jurisdictions develop and sustain capabilities for preventing, protecting against, responding to, and recovering from threats or acts of terrorism. The FY 2009 UASI program is specifically intended to enhance regional preparedness efforts. Urban areas must use these funds to implement regional approaches to overall preparedness and are encouraged to adopt regional response structures whenever appropriate.

In 2008, DHS created the UASI Nonprofit Security Grant Program, in part to integrate nonprofit preparedness activities with broader state and local preparedness efforts.

Citizen Corps. The Citizen Corps focuses on education, training, and volunteer service to make communities safer and better prepared to respond to disasters, including public health issues and terrorist threats. Local Citizen Corps Councils guide local efforts, including the CERT program (supported by FEMA), and the Medical Reserve Corps (see next section on HHS).

CERT. The CERT program was originally developed by the Los Angeles Fire Department in 1995. CERT's main goal is to educate residents about emergency preparedness. CERT members are also trained by first responders in basic principles of emergency preparedness, then assist their community members when an event happens. In particular, the CERT serves as a bridge for education and response when professional first responders are not immediately available. FEMA supports CERTs by providing or sponsoring train-the-trainer courses around the country.

MMRS. The purpose of the MMRS is to foster an integrated and coordinated approach to medical response planning and operations. FEMA established the MMRS in 1996, partly in response to the 1995 sarin gas attacks in Tokyo and the Oklahoma City bombing. The MMRS works with all entities in a jurisdiction, including emergency management, medical, public health, law enforcement, fire, and EMS, to develop plans, conduct exercises, and organize medical resources to enhance a jurisdiction's ability to respond to a potential mass casualty event. The MMRS is particularly critical in the first hour of response, when reducing casualties and implementing life-saving techniques are critical. There are currently 124 MMRS programs across the country, supporting the most populous jurisdictions.

U.S. Department of Health and Human Services

HHS has primarily focused on providing medical support, including countermeasures, in disaster events and developing guidance for medical treatment in mass casualty events. Four entities in HHS have major responsibilities: the Assistant Secretary for Preparedness and Response (ASPR), the Office of the Surgeon General, CDC, and AHRQ. Organizations and programs in each of the entities are focused on specific aspects of medical and public health preparedness and response to major disasters and other emergencies.

ASPR. ASPR is the principal adviser to the Secretary of HHS "on matters related to bioterrorism and other public health emergencies." ASPR is also responsible for related interagency coordination with other federal departments, agencies, and offices, and state and local officials on matters involving bioterrorism and other public health emergencies. For more information, see ASPR (undated).

Hospital Preparedness Program (HPP). HPP, a subordinate entity of ASPR—supports the health-care service elements relevant to public health emergencies. Specifically, the program aims to improve the capabilities of hospitals and health-care systems—including inpatient, outpatient, and other relevant facilities—to prepare for and respond to bioterrorism or other public health emergencies. The program takes an all-hazards approach and currently focuses on strengthening selected priority capabilities: interoperable communications systems, bed tracking, personnel management, fatality-management planning, and hospital-evacuation planning

NDMS. NDMS—another subordinate entity of ASPR for the HHS-led functions—is a federally coordinated system comprising HHS-, VA-, and DoD-led efforts to augment and supplement federal, state, and local medical response to major disasters and other emergencies (including natural disasters, major transportation accidents, technological disasters, and acts of terrorism). NDMS components can provide medical response (HHS led; personnel, supplies, and equipment); patient transport (DoD led); and medical treatment at hospitals outside of an affected area (VA led).

Office of the Surgeon General. The U.S. Surgeon General, based in the Office of the HHS Secretary, oversees the 6,000-member Commissioned Corps of the U.S. Public Health Service and provides health education to the American public on how to improve health and reduce risk of illness and injury. The Office of the Surgeon General also oversees the Medical Reserve Corps (MRC).

MRC. The MRC, a partner program with Citizen Corps, is a network of volunteers, designed to supplement local emergency and public health capabilities. MRC volunteers, who donate their time and expertise, include "medical and public health professionals such as physicians, nurses, pharmacists, dentists, veterinarians, and epidemiologists." For more information, see MRC (2009).

CDC. CDC is one of 11 HHS operating divisions, responsible for public health surveillance and prevention and control of diseases and injuries. CDC strengthens existing public health infrastructure while working with partners throughout the nation and the world. CDC supports state and local public health emergency preparedness through three key initiatives.

Cooperative Agreements on Public Health Emergency Preparedness. The Public Health Emergency Preparedness cooperative agreements provide funding to enable public health departments to develop the capacity and capabilities to prepare for and respond effectively to emergencies, such as natural disasters and emerging infectious diseases—including bioterrorism incidents. These response efforts are designed to support the NPG, the NRF, and the NIMS, as operationalized by state and local health departments.

SNS. The SNS is the federal government's largest national repository of medicine and medical supplies, which can be deployed rapidly to supplement local supplies to protect the public against naturally occurring and intentional threats to health. The SNS contains antibiotics, chemical antidotes, antitoxins, life-support medications, intravenous drug administration, airway-maintenance supplies, and medical and surgical items. For more information about SNS assets and how they are deployed, see CDC (2009).

CRI. CRI is designed to help the largest U.S. cities and metropolitan areas develop the ability to provide countermeasures in a large-scale public health emergency, including a bioterrorist attack or disease outbreak. The primary goal of CRI is to help jurisdictions plan to provide antibiotics to their entire population within 48 hours of an event. At this writing, CRI includes 72 metropolitan regions, representing 57 percent of the U.S. population. A current listing of CRI areas can be found at CDC (2008).

AHRQ. AHRQ is also one of 11 HHS operating divisions, responsible for health-services research. It provides a number of public health emergency management resources, including tools and information related to community planning, mass prophylaxis, modeling, pandemic influenza, pediatrics, and surge capacity; evidence reports; and notes from selected meetings and conferences. Two documents from the list of tools and resources are highlighted here, as relevant examples of the types of information provided by AHRQ.

***Altered Standards of Care in Mass Casualty Events* (AHRQ, 2005a).** This guidance examines how current standards of care would need to be altered in response to a mass casualty event, identifies what tools are needed to ensure an effective health and medical-care response, and recommends action to address the needs of federal, state, regional, and community planners. In particular, the guidance describes the need to examine standards of care related to (1) scope of practice, (2) patient privacy, (3) facility standards, and (4) guidelines regarding use of other qualified responders, including reserve military medical and nursing providers and modifying state licensing for out-of-state providers on a temporary basis.

***Mass Medical Care with Scarce Resources: A Community Planning Guide* (AHRQ, 2007a).** A follow-on document to *Altered Standards of Care in Mass Casualty Events*, this guide is intended to outline key issues when planning for a mass casualty event, particularly with respect to legal issues, setting up alternative care sites, and identifying resources for palliative care. The guide argues that planning a public health response to a mass casualty event must be comprehensive, community based, and coordinated at the regional level, including developing robust security plans that incorporate uniformed personnel (e.g., National Guard).

U.S. Department of Justice

JTTFs. The JTTFs, a multiagency effort led by the U.S. Department of Justice (DOJ) and the FBI, comprise investigators, analysts, linguists, SWAT team experts, and other specialists from dozens of U.S. law enforcement and intelligence agencies. JTTF members track down leads, gather evidence, make arrests, provide security for special events, conduct training, collect and share intelligence, and respond to threats and incidents. About 60 percent of the full-time JTTF members are FBI special agents; a little over 22 percent are representatives of other federal agencies, and the remainder are from state and local entities. JTTFs operate in 100 cities nationwide.

Anti-Terrorism Advisory Councils (ATACs). The goal of the ATACs is to provide more comprehensive and better-coordinated prevention and prosecution operations. The councils comprise federal, state, and local law enforcement, public health and safety officials, and, in some cases, private-sector members. ATACs exist in each of the 93 U.S. Attorneys Offices.

State-Level Mechanisms

Mutual Assistance

Many states have taken significant steps to improve preparedness, response, and recovery capabilities. All 50 states, as well as the District of Columbia, Puerto Rico, and the U.S. Virgin Islands, have enacted legislation to become part of EMAC, a mutual aid compact that facilitates resource sharing across state lines during times of disaster and emergency.

EMAC has been used extensively in major disasters. For example, during Hurricanes Katrina and Rita, 48 states provided support through EMAC to stricken areas of other states. EMAC has also been implemented more recently for wildland fires in California in 2007 and 2008 and the major floods in the Midwest in 2008. The national EMAC system does not allow localities to use that system directly—all support and assistance for localities must be coordinated and agreed upon at the state level.

At the local level, there is no standard, national system or process for mutual assistance. Some localities have—by virtue of proximity or similar threats—entered into agreements for mutual support, but the scope of these agreements varies significantly.

National Guard Organizations

Several states have established organizations within the National Guard structure that have been supported with direct appropriations from Congress to provide assistance at the state and local levels for various types of incidents. In this section, we describe several of these that are especially relevant to our study.

CSTs. CSTs are designed to deploy rapidly and provide initial assessments of incidents involving CBRN agents and to provide robust communications to obtain other military support if requested by civilian authorities. There are 56 CSTs—one in each state (except California and Florida, each of which has two), and one each in the District of Columbia, Puerto Rico, the U.S. Virgin Islands, and Guam. Although federally funded, the CSTs are under the direct control of each state or territory's governor. Each team comprises 22 full-time Army and Air National Guard personnel, and a suite of communications and analysis equipment.

Although intended for support in major CBRN events, CSTs have, however, been employed following a variety of other major incidents, including the 9/11 attacks, Hurricane Katrina response, and the recovery of the space shuttle Columbia. They have also been deployed in connection with several National Special Security Events, such as the Super Bowl, the XIX Olympic Winter Games in Salt Lake City, presidential inaugurations, and national political-party conventions. CSTs are available to governors for deployment out of the home state under the provisions of EMAC.

CBRNE Enhanced Response Force Package (CERFP). The CERFP is designed to provide immediate response capabilities to a governor for search and rescue operations, decontamination, and medical triage and initial treatment. Each CERFP is staffed by personnel from already established National Guard units. There is at least one CERFP in each of the 10 FEMA regions. As of March 2009, there were 12 validated CERFPs, and an additional five CERFPs have been authorized and funded by Congress. As with CSTs, a CERFP may be deployed to an incident outside of its state, under the provisions of EMAC.

APPENDIX C

U.S. Department of Defense Policy, Doctrine, and Relevant Organizations

OSD and Joint Policy and Doctrine

DoD Directive 1100.20, Support and Services for Eligible Organizations and Activities Outside the Department of Defense, April 12, 2004

Units and personnel of the armed forces are authorized, on request, to assist certain eligible organizations and activities in addressing community and civic needs in U.S. states and territories, when such assistance is incidental to military training or is otherwise authorized by law. The purpose is to build on the long-standing tradition of the armed forces of the United States, acting as good neighbors at the local level, in applying military personnel to assist worthy civic and community needs. Activities authorized by the directive include health-care services, general engineering, and critical infrastructure protection.

DoD Directive 2000.12, DoD Antiterrorism (AT) Program, August 18, 2003

This directive provides policy guidance for protecting DoD components and personnel from terrorist acts by establishing a high-priority, comprehensive antiterrorism program. Commanders at all levels have the responsibility and authority to enforce appropriate security measures and maintain antiterrorism awareness and readiness in order to protect DoD elements and assigned or attached personnel subject to their control (including dependent family members).

DoD Directive 2060.02, Department of Defense (DoD) Combating Weapons of Mass Destruction (WMD) Policy, April 19, 2007

This directive provides DoD policy to combat WMD by measures that will dissuade, deter, and defeat those who use or threaten to use WMD to harm the United States, its citizens, its armed forces, and its friends and allies. It is designed to support the National Strategy to Combat Weapons of Mass Destruction (White House Office, 2002), to include supporting military force planning and doctrine to organize, train, exercise, and equip military forces to combat WMD. DoD will also maintain the ability to respond to and mitigate the effects of WMD use. Among DoD's combating-WMD strategic goals is the imperative of U.S. armed forces to protect against, respond to, and recover from WMD use. Consequence management is one of the eight combating-WMD mission areas supported by DoD.

DoD Directive 3003.01, DoD Support to Civil Search and Rescue (SAR), January 20, 2006

This directive provides guidance for supporting domestic civil authorities providing civil SAR service to the fullest extent practicable on a noninterference basis with primary military duties. It is intended to be consistent with other applicable national directives, authorities, plans, guidelines, and agreements, and generally on a reimbursable basis according to the Economy Act (Pub. L. 73-2, 1933) or the Stafford Act. To ensure a coordinated DoD response, all civil SAR operations conducted in support of domestic civil authorities following a presidential declaration of a disaster or emergency or in response to incidents are designated as incidents of national significance by the Secretary of Homeland Security.

DoD Directive 3025.1, Military Support to Civil Authorities (MSCA), January 15, 1993

This directive provides policy guidance to all major DoD organizations for support to civil authorities for most domestic emergencies and disasters, including the use of the military under the provisions of the Stafford Act. It specifically excludes from its scope military assistance for civil law enforcement operations. This directive is significantly dated and is currently undergoing revision and consolidation with related directives. Nevertheless, it specifies DoD components' obligations to respond to civil authorities by consolidating previously existing policy and responsibilities applicable to disaster-related civil emergencies within the United States and its territories. Significant excerpts follow to describe DoD civil support policy:

DoD Directive 3025.12, Military Assistance for Civil Disturbances (MACDIS), February 4, 1994

This directive provides specific guidance for the implementation by DoD entities of the provisions of the Insurrection Act (10 U.S.C. 331–335). It provides for DoD support to civilian law enforcement agencies, at the direction of the president, to protect life and property and to maintain law and order in the civilian community during insurrections, rebellions, and other types of domestic violence.

DoD Directive 3025.15, Military Assistance to Civil Authorities, February 18, 1997

This directive provides DoD policy guidance for activities to cooperate with and provide military assistance to civil authorities as directed by and consistent with applicable law, presidential directives, executive orders, and this directive. Pursuant to this directive, all requests by civil authorities for DoD military assistance are evaluated by DoD approval authorities against the following criteria: (1) legality, (2) potential lethality or use of force by or against DoD forces, (3) risk and safety of DoD forces, (4) cost (i.e., impact on DoD budget and reimbursement), (5) appropriateness of the mission to be conducted by DoD forces, and (6) impact on DoD force readiness. Civil requests for immediate response may go directly to a DoD component or military commander to save lives, prevent human suffering, or mitigate great property damage under imminently serious conditions. The DoD components that receive verbal requests from civil authorities for support in an emergency may initiate informal planning and, if required, immediately respond as authorized in DoDD 3025.1. Civil authorities must follow any verbal requests for emergency support with a written request.

DoD Directive 3025.16, Military Emergency Preparedness Liaison Officer (EPLO) Program, December 18, 2000

This directive provides guidance to military departments for representatives of the federal military in each state and in each of FEMA's ten regional offices. EPLOs coordinate the input of military personnel, equipment, and supplies to support the emergency relief and cleanup efforts of civil authorities (as described in DoDDs 3025.1, 3025.12, and similar). It establishes

DoD oversight for the EPLO programs in the areas of staffing, readiness, equipping, training, funding, and exercises.

DoD Directive 3160.01, Homeland Defense Activities Conducted by the National Guard, August 25, 2008

This directive provides for the payment of certain activities of the National Guard, remaining under the command and control of state governors, for homeland defense operations requested by a governor and that the Secretary determines to be necessary and appropriate. Activities funded by DoD may include "deliberate, planned activities" or for "exceptional circumstances."

DoD Directive 5525.5, DoD Cooperation with Civilian Law Enforcement Officials, January 15, 1986

This directive provides policy guidance for support to law enforcement officials not included under DoDD 3025.12, such as counterdrug operations. That guidance includes procedures for providing any information collected during the course of normal military operations to civilian law enforcement agencies "that may be relevant to a violation of any federal or state law within the jurisdiction of such officials." It also provides procedures for making DoD equipment and facilities available to civilian law enforcement entities. While specifically noting the restrictions contained in the Posse Comitatus Act, it acknowledges the authority of DoD entities to engage in law enforcement activities under the Insurrection Act and other statutes discussed elsewhere.

DoD Directive 6200.3, Emergency Health Powers on Military Installations, May 12, 2003

This directive establishes DoD policy under applicable law to protect installations, facilities, and personnel in the event of a public health emergency due to biological warfare, terrorism, or other public health emergency communicable disease epidemics. The policy defines and addresses public health emergency, communicable disease, quarantinable communicable disease, isolation, quarantine, and temporary restriction of movement. It is DoD policy that military installations, property, and personnel and other individuals working, residing, or visiting military installations shall be protected under applicable legal authorities against communicable diseases associated with biological warfare, terrorism, or other public health emergencies. Notably, this policy specifies the requirements for the installation-level PHEO. The policy states that every military commander who is required by DoDI 5200.08, Security of DoD Installations and Resources, to issue regulations for protecting and securing property or places under his or her command, shall designate a PHEO. The installation PHEO will be a senior health-profession military officer or DoD civilian employee affiliated with the command of the commander or of a higher-level or associated command. Additionally, every health-care provider on an installation (military, government civilian, or contractor), every pharmacist, and every veterinarian will report to the appropriate PHEO any circumstance suggesting a public health emergency (in addition to requirements to report such circumstances to applicable non-DoD disease surveillance and reporting systems). PHEOs are empowered with broad responsibilities to identify and monitor events that may affect public health, monitor for disease outbreaks, and recommend actions to command authorities to mitigate risk of disease occurrence or spread. PHEOs are also required to coordinate with applicable federal, state, and local public health and law enforcement officials to protect public health and safety. Commanders, as identified in DoDI 5200.08, may declare a public health emergency on one or more military installations under his or her command. Upon such a declaration, the commander, in consultation with the PHEO, may implement emergency powers specified in this directive and must make an immediate report through the chain of command to the Secretary of Defense and the appropriate service surgeon general or combatant command surgeon. A commander may exercise a broad range of emergency and special powers inherent in this policy (examples include but are not limited to testing of individuals for disease, evacuation of facilities, control of evacuation routes, restriction of movement, and quarantine). Such powers may include persons other than military personnel who are present on a DoD installation or other area under DoD control. Finally, the policy specifies that protected health information shall be used and disclosed as necessary to ensure proper treatment of individuals and to prevent the spread of communicable diseases.

DoD Instruction 2000.16, DoD Antiterrorism Standards, October 2, 2006

This DoD instruction updates policy implementation, responsibilities, and the antiterrorism standards for DoD components. This instruction states DoD's policy to protect DoD personnel, their families, installations, facilities, information, and other material resources from terrorist acts and to establish antiterrorism standards for DoD. The instruction states that commanders at all levels shall have the authority to enforce security measures and are responsible for protecting persons and property subject to their control and that geographic combatant commander antiterrorism policies and programs shall take precedence over all other antiterrorism policies (with limited exceptions outside the continental United States [OCONUS]). Non-DoD tenants on a DoD installation, facility, or other DoD property must comply with all aspects of the DoD antiterrorism program addressed in this instruction and other antiterrorism guidance documents.

DoD Instruction 2000.18, Department of Defense Installation Chemical, Biological, Radiological, Nuclear, and High Yield Explosive Emergency Response Guidelines, December 4, 2004

This instruction implements and assigns responsibilities on DoD policy response to CBRNE events. Commanders and those responsible for CBRNE response must be prepared to respond to CBRNE events in order to save life, prevent human suffering, mitigate the incident, and protect personnel and resources affected by such events.

DoD Instruction 5200.08, Security of DoD Installations and Resources, December 10, 2005

This instruction provides general guidance for protection of DoD installations, property, and personnel. Commanders at all levels have the responsibility and authority to enforce appropriate security measures to ensure the protection of DoD property and personnel assigned, attached, or subject to their control. DoD commanders have the authority to take reasonably necessary and lawful measures to meet these requirements. The following commanders are specifically identified: (1) the commanding officers of all military reservations, posts, camps, stations, or installations subject to the jurisdiction, administration, or in custody of the Department of the Army; (2) the commanding officers of all naval ships or afloat units, commanders, or commanding officers of naval shore activities or installations, bases, camps, stations, and supply activities, subject to jurisdictions, administrations, or in custody of the Department of the Navy; (3) the commanders of major air commands, numbered air forces, wings, groups, or installations subject to the jurisdiction, administration, or in custody of the Department of the Air Force; (4) the commanders of installations or activities subject to the jurisdiction, administration, or in custody of the defense agencies, DoD field activities, and other DoD organizational entities, or their operating activities; (5) the commanders of installations or activities subject to the jurisdiction, the administration, or in the custody of the commanders of the combatant commands, or the Chairman of the Joint Chiefs of Staff; and (6) the commanders and civilian directors in the chain of command immediately above an installation or activity not headed by a commander or civilian equivalent who issues regulations or orders on the security of the installation or activity.

Joint Publication (JP) 3-27, Homeland Defense, July 12, 2007

This publication provides doctrine for the defense of the U.S. homeland across the range of military operations. It provides information on command and control, interagency and multinational coordination, and operations required to defeat external threats to, and aggression against, the homeland. It sets forth joint doctrine to govern the activities and performance of the armed forces of the United States in operations and provides the doctrinal basis for interagency coordination for defense of the homeland. It provides military guidance for the exercise of authority by combatant commanders and other joint force commanders (JFCs) and prescribes joint doctrine for operations and training. It provides military guidance for use by the armed forces in preparing their appropriate plans. It is not the intent of this publication to restrict the authority of the JFC from organizing the force and executing the mission in a manner the JFC deems most appropriate to ensure unity of effort in the accomplishment of the overall objective.

JP 3-28, Civil Support, September 14, 2007

This publication provides overarching guidelines and principles to assist commanders and their staffs in planning and conducting joint civil support operations. It sets forth joint doctrine to govern the activities and performance of the armed forces of the United States in civil support operations and provides the doctrinal basis for interagency coordination during domestic civil support operations. It provides military guidance for the exercise of authority by combatant commanders and other JFCs and prescribes joint doctrine for operations, education, and training. It provides military guidance for use by the armed forces in preparing their appropriate plans. It is not the intent of this publication to restrict the JFC's authority from organizing the force and executing the mission in a manner the JFC deems most appropriate to ensure unity of effort in the accomplishment of the overall objective.

Relevant Military Department Policy and Doctrine

U.S. Department of the Army

Army Regulation (AR) 525-13, Antiterrorism, September 11, 2008. This regulation prescribes policy, procedures, standards, and guidance, and assigns responsibilities for the Army antiterrorism program, implementing DoDD 2000.12 and DoDI 2000.16. It includes significant and specific requirements for coordination, planning, and other activities with local civilian authorities. It states that installation commanders will incorporate antiterrorism into their overall force protection (FP) program; establish antiterrorism risk management procedures (threat, vulnerability, and consequence assessments) appoint an antiterrorism officer; establish an antiterrorism working group, a threat working group (to include local, state, and federal law enforcement) and an antiterrorism executive committee; publish guidance for the execution of antiterrorism standards within the overarching FP security program; designate a focal point to coordinate and act on information received from federal, state, local, host-nation, and other intelligence agencies; ensure that all tenant and supported reserve component (RC) units and activities are participants in the antiterrorism planning process and are included in antiterrorism plans; and implement and execute Army antiterrorism standards in accordance with implementing guidance in the regulation. Commanders will establish working relationships with local civilian communities to defend against terrorism, including coordination of antiterrorism plans with local officials to ensure understanding of what military or civilian support would be rendered in the event of a major incident. Commanders must also establish MAAs with local authorities to facilitate the shared use of critical resources and to coordinate security measures and assistance requirements to ensure the protection of Army personnel and their family members at off-installation facilities and activities.

U.S. Army Medical Command (MEDCOM) Regulation 525-4, U.S. Army Medical Command Emergency Management, December 11, 2000. This regulation requires that MEDCOM and its subordinate activities (commands) be prepared to respond to crises, either foreign or domestic. MEDCOM will operate in accordance with DoD's concept of operations inherent in DoD policy and the NRF. MEDCOM's major medical assets are delineated in this regulation. Emergency planners at all levels will develop and maintain emergency response plans in accordance with this regulation.

U.S. Army Medical Command Pamphlet 525-1, Medical Emergency Management Planning, October 1, 2003. This publication amplifies the policy set forth in MEDCOM regulation 525-4, compiling emergency management and response information into a single source document for Army medical planners. The document provides MEDCOM guidance on homeland security and responses to emergencies, disasters, and incidents involving WMD or CBRNE devices. The document provides a strategic overview of emergency management planning as well as detailed operational and tactical concepts and procedures for emergency planners. The pamphlet directs an all-hazards approach for medical emergencies; plans generated within MEDCOM organizations will be designed for response to any medical emergency, taking into account all potential threats and vulnerabilities in medical contingency planning. The pamphlet also states that MEDCOM medical treatment, disease prevention, and research facilities are local resources, expected to participate as members of local communities' immediate responses when resident in a disaster area. Because the federal government provides personnel, equipment, supplies, facilities, and managerial, technical, and advisory services when supporting a local or state government that has been overwhelmed, it follows that MEDCOM assets could become part of this disaster response framework for providing assistance after a disaster or emergency.

U.S. Department of the Navy

Office of the Chief of Naval Operations (OPNAV) Instruction 3440.17, (OPNAV 3440.17), Navy Installation Emergency Management (EM) Program, July 22, 2005. This OPNAV instruction provides policy, guidance, operational structure, and assignment of responsibilities for a comprehensive, all-hazards emergency management program at Navy regions and installations. It specifies certain required capabilities for various incidents. This instruction establishes CNI responsibility and authority to develop, implement, and sustain a comprehensive emergency management program at regions and installations capable of effective all-hazards preparedness, mitigation, response, and recovery, in order to save lives, protect property, and sustain mission readiness. It requires that naval installations' EMPs be consistent with state, local, and other-service emergency management or contingency plans and that all-hazards consequence, threat, and hazard assessments be integrated with those of adjacent and nearby federal, state, local, other-service, and private agencies and departments to the greatest extent possible. It states that installation commanders should seek to participate in federal, state, local, other-service, and private emergency management planning, training, and exercises, and that commanders should encourage reciprocal participation by these entities in regional and installation emergency management planning, training, and exercises.

OPNAV Instruction 3440.16c, Naval Civil Emergency Management Program, March 10, 1995. This instruction states that the primary objective of the Navy emergency management program is to protect and restore Navy mission capabilities. However, Navy commanders at all levels must be prepared to employ appropriate Navy resources (personnel, forces, equipment, supplies, and facilities) in support of civil authorities for certain emergencies (specifically referencing the DoD directives on support to civil authorities). This instruction carries forward the authority for installation commanders to conduct immediate-response activities if requested by local civilian authorities. Navy activity commanders will support, within capabilities, host-base civil-assistance programs as outlined in host-tenant agreements, MOUs, or host-base plans or instructions. Support includes required coordination with and participation in emergency planning and exercises with local and state entities.

CNI Instruction 3440.17, Navy Installation Emergency Management Program Manual, January 23, 2006. This manual provides additional detail for implementation of the Navy installation emergency management program required in OPNAV Instruction 3440.17. Notably, at the installation level, it requires that the installation commander establish an installation emergency management working group (EMWG) based on the standards set forth in the document. The primary purpose of the installation EMWG is the development and coordination of the installation emergency management plan across participating departments and offices. The installation EMWG shall also assist the installation emergency management officer in the categorization of personnel, the inventory of response and recovery resources currently and potentially available to the installation, the conduct of a hazard assessment, and assessment of the current response capabilities of the installation.

U.S. Department of the Air Force

Air Force Instruction (AFI) 10-2501, Air Force Emergency Program Planning and Operations, January 24, 2007. This instruction addresses the primary mission of the Air Force Emergency Management (EM) program: to save lives, minimize the loss or degradation of resources, and continue, sustain, and restore operational capabilities in an all-hazards physical threat environment at Air Force installations worldwide. The instruction also states that the ancillary missions of the Air Force emergency management program are to support homeland defense and civil support operations and to provide support to civil and host-nation authorities in accordance with DoDDs and through the appropriate combatant commands. The instruction states that, at the installation level, the host wing's Civil Engineering Squadron (Readiness Flight) is the office of primary responsibility for the installation commander's emergency management program. In developing the emergency management program, objectives and program elements related to all-hazards threats must be addressed.

AFI 10-802, Military Support to Civil Authorities, April 19, 2002. Under this AFI, the Air Force Deputy Chief of Staff Air and Space Operations (AF/A3) is designated as the principal planning agent for Air Force homeland security activities. AF/A3 has delegated authority for planning and training and facilitating the execution of Air Force military support to civil authorities (MSCA) support to the Air Force Director of Homeland Security, within the AF/A3 directorate. The Air Force Director of Homeland Security performs all functions as required by DoDD 3025.1, Military Support to Civil Authorities.

AFI 41-106, Unit Level Management of Medical Readiness Programs, April 14, 2008, incorporating through change 2, July 28, 2009. This instruction sets procedures for medical readiness planning and training for wartime, humanitarian assistance, homeland security and defense, and disaster response contingencies. The instruction sets the requirements for installation-level planners, including disaster response. The medical-contingency readiness plan (MCRP) is developed and updated annually by the installation-level medical group (readiness plans and operations flight). Relevant annexes to the MCRP are Annex B (Medical Group Commander and Medical Control Center), Annex C (Patient Redistribution), Annex D (Casualty Management), Annex E (Public Health Team), Annex F (Bioenvironmental Engineering Team), Annex G (Medical Logistics), Annex M (Civilian Disturbances), Annex N (Terrorist and WMD Threats), and Annex T (Disaster Response and Recovery). Additionally, an Air Force medical group's readiness plans and operations flight will prepare or coordinate inputs into the installation support plan.

DoD Organizations with Homeland Defense and Civil Support Responsibilities

DoD now has numerous organizations that have homeland defense and civil support as either primary or secondary missions. Several of the ones relevant to this report are described here.

Assistant Secretary of Defense for Homeland Defense and Americas' Security Affairs (ASD [HD&ASA])

Within DoD, policy for homeland defense falls under the purview of the ASD (HD&ASA). In December 2002, the president signed into law the Bob Stump National Defense Authorization Act for Fiscal Year 2003 (Pub. L. 107-314), with Section 902 stating, "One of the Assistant Secretaries (of Defense) shall be the ASD (HD). He shall have as his principal duty the overall supervision of the homeland defense activities of the Department of Defense."[1] The ASD falls under the Under Secretary of Defense for Policy (USD [P]) and is responsible for assisting the Secretary of Defense in policy direction on homeland defense matters. In coordination with the Chairman of the Joint Chiefs of Staff, the ASD provides oversight to DoD homeland defense activities, develops policies, conducts analyses, provides advice, and makes recommendations on homeland defense, DSCA, emergency preparedness, and domestic-crisis management matters. Such recommendations may be coordinated through the Chairman of the Joint Chiefs of Staff to U.S. Northern Command and other combatant commands to guide their planning and execution activities. The ASD also serves as the DoD domestic-crisis

[1] The Secretary of Defense recently expanded the title and placed within the ASD's portfolio the Office of Western Hemisphere Affairs.

manager and represents DoD on all homeland matters with designated lead federal agencies, the Executive Office of the President, the Department of Homeland Security, other executive departments and federal agencies, and state and local entities, as appropriate.

Assistant Secretary of Defense for Health Affairs (ASD [HA])

The ASD (HA) falls under the Under Secretary of Defense for Personnel and Readiness (USD [P&R]). ASD (HA) is the principal staff assistant and adviser to USD (P&R) and the Secretary and Deputy Secretary of Defense for all DoD health policies, programs, and activities. ASD (HA) is charged with executing DoD's medical mission, which is to provide and maintain readiness to provide medical services and support to members of the armed forces during military operations and to provide medical services and support to members of the armed forces, their dependents, all other beneficiaries entitled to DoD medical care under Title 10 of the U.S. Code. Per DoDD 2060.02, Department of Defense (DoD) Combating Weapons of Mass Destruction (WMD) Policy, the USD (P&R) will establish procedures and standards and ensure implementation of "DoD Combating WMD Force Health (FH) Protection" policy, training, and readiness. To fulfill these responsibilities, ASD (HA) is the USD (P&R) point of contact for combating WMD. Within the Office of the ASD (HA), the Deputy Assistant Secretary of Defense for Force Health Protection and Readiness oversees DoD efforts to develop and implement policies and programs relating to deployment medicine, force health protection, national disaster support, and medical readiness.

Chairman, Joint Chiefs of Staff (CJCS)

CJCS is the senior military adviser to the President of the United States and Secretary of Defense. CJCS advises the President and Secretary of Defense on operational policies, responsibilities, and programs and assists the Secretary of Defense in implementing operational responses to threats or acts of terrorism. CJCS also translates Secretary of Defense guidance into operation orders to provide assistance to a lead federal agency. For civil support, CJCS is the principal military adviser to the President and Secretary of Defense in preparing for and responding to CBRNE incidents, ensuring that military planning is accomplished to support a lead federal agency for crisis or consequence management activities. Other specific responsibilities with respect to homeland defense are found in JP 3-26.

U.S. Northern Command (USNORTHCOM)

Commander, U.S. Northern Command, has specific responsibilities for homeland defense and for supporting civil authorities within the assigned area of responsibility. USNORTHCOM conducts operations to deter, prevent, and defeat threats and aggression aimed at the United States, its territories, and its interests within the assigned area of responsibility. As directed by the President or Secretary of Defense, USNORTHCOM provides military assistance to civil authorities, including consequence-management operations. The following components fall under USNORTHCOM:

Joint Force Headquarters National Capital Region (JFHQ-NCR). After September 11, 2001, JFHQ-NCR was established as the responsible headquarters for land-based homeland defense, DSCA, and incident management in the National Capital Region (District of Columbia and neighboring counties and cities in Maryland and Virginia) in a supporting role to a lead federal agency. JFHQ-NCR's mission is to "plan, coordinate, maintain situational awareness, and as directed, employ forces for homeland defense and military assistance to civil authorities in the National Capital Region Joint Operations Area to safeguard the Nation's capital" (JFHQ-NCR, undated). JFHQ-NCR provides a continuous ability to support national security requirements in response to emergencies or National Special Security Events. Once an event is designated, the command becomes Joint Task Force—National Capital Region, which would then direct military assistance to federal and other civil authorities. (For more information, see JFHQ-NCR, undated.)

Joint Task Force Civil Support (JTF-CS). JTF-CS is an established standing Joint Task Force headquarters, currently located on Fort Monroe, Virginia. JTF-CS is charged to assist civil authorities in conducting CBRNE countermeasures within the USNORTHCOM area of responsibility. Because DoD is only one member of the federal response community, JTF-CS is in constant coordination with other federal, state, and local agencies. Whereas the most likely supported agency during an incident of national significance would be FEMA (which falls under DHS), JTF-CS also maintains liaison with other key organizations, including DOJ, the Department of Energy, the CDC, and various state and local emergency, law enforcement, medical, and public health agencies, and state National Guard headquarters. (For more information, see JTF-CS, undated.)

Joint Task Force Alaska (JTF-AK). JTF-AK is a component of USNORTHCOM headquartered at Elmendorf AFB, Alaska. JTF-AK coordinates the land defense of Alaska and plans for defense support to Alaska's civil authorities and to federal lead agencies, such as FEMA. JTF-AK's civil support mission includes domestic disaster relief operations that occur following fires, hurricanes, floods, and earthquakes. Support also includes managing the consequences of a WMD terrorist event. (For more information, see USNORTHCOM, undated [a].)

Chemical, Biological, Radiological, Nuclear, and High-Yield Explosives Consequence Management Response Force (CCMRF). The CCMRF is a team of about 4,700 joint personnel that would deploy as DoD's initial response force for a CBRNE incident. Its capabilities include SAR, decontamination, medical, aviation, communications, and logistical support.

Each CCMRF is composed of three functional task forces—Task Force Operations, Task Force Medical, and Task Force Aviation—that have their own individual operational focus and set of mission skills. Depending on the different mission requirements and the incident commander's priorities, Task Force Operations, Task Force Medical, and Task Force Aviation units would have varying roles and responsibilities based on the type of catastrophe and the size of the geographical area. In USNORTHCOM's first assigned CCMRF, the Army's 3rd Infantry Division's 1st Brigade Combat Team, assigned at Fort Stewart, Georgia, forms the core unit of Task Force Operations.

Although CCMRFs are a joint force comprised of soldiers, sailors, airmen, and marines, the first CCMRF falls under the operational control of USNORTHCOM's Joint Force Land Component Command, U.S. Army North, located in San Antonio, Texas. JTF-CS, USNORTHCOM's subordinate command in Fort Monroe, Virginia, would serve as the operational headquarters and work closely with state and local officials and first responders. (For more information, see USNORTHCOM, undated [b].)

U.S. Pacific Command (USPACOM)

USPACOM has responsibility for conducting U.S. homeland defense and civil support activities, similar to those of USNORTHCOM, within most of its area of responsibility. That area of responsibility includes the state of Hawaii and U.S. insular areas in the Pacific. As noted earlier, despite the fact that USPACOM has a major subordinate command in Alaska

(Alaska Command), the homeland defense and civil support entity—JTF-AK—is assigned to USNORTHCOM. The following component of USPACOM is relevant to this study:

Joint Interagency Task Force West (JIATF West)

JIATF West has been providing support to the U.S. counterdrug effort since 1989, now focused exclusively on Asia and the Pacific. Its mission is

> to conduct activities to detect, disrupt and dismantle drug-related transnational threats in Asia and the Pacific by providing interagency intelligence fusion, supporting U.S. law enforcement, and developing partner nation capacity in order to protect U.S. security interests at home and abroad. (USPACOM, 2008)

The JIATF West staff includes both uniformed and civilian members of all five military services, as well as representatives from the national intelligence community and U.S. federal law enforcement agencies. Law enforcement representatives include the DEA, FBI, and U.S. Immigration and Customs Enforcement (ICE). JIATF West is closely aligned with USPACOM's theater security cooperation, counterterrorism, and maritime security priorities in planning, developing, and implementing its counterdrug programs. (For more information, see USPACOM, 2008.)

Other Specialized Units

National Guard Organizations. See Appendix B.

U.S. Marine Corps Chemical Biological Incident Response Force (CBIRF) The Marine Corps' CBIRF supports local, state, and federal agencies and unified combatant commanders in implementing consequence-management operations in response to credible threats of a CBRNE incident. CBIRF provides capabilities for agent detection and identification; casualty search, rescue, and personnel decontamination; and emergency medical care and stabilization of contaminated personnel. (For more information, see USMC, undated.)

U.S. Coast Guard. The Coast Guard has command responsibilities for the U.S. Maritime Defense Zone, countering potential threats to U.S. coasts, ports, and inland waterways through port-security, harbor-defense, and coastal-warfare operations and exercises. By statute, the Coast Guard is an armed force, operating in the joint arena at any time and functioning as a specialized service under the U.S. Navy in time of war or when directed by the President. Organized under DHS, the U.S. Coast Guard provides military, humanitarian, and civilian law enforcement capabilities.

Military Health System (MHS). The MHS is not a DoD component, nor is it a single entity. Instead, *MHS* is a term that describes the collective set of medical and dental programs, personnel, facilities, and other assets operating under the policy purview of the ASD (HA) and the administrative control (day-to-day operations; organize, train, and equip responsibilities) of each respective military department pursuant to Chapter 55 of Title 10 of the U.S. Code, by which DoD provides health-care services and support to the armed forces during military operations, and health-care services and support under TRICARE to members of the armed forces, their family members, and other beneficiaries entitled to DoD medical care. Recent history provides examples of MHS assets being tasked through appropriate command channels to support homeland defense activities and domestic contingency response.

APPENDIX D

Site Visit Interview Protocols and Synthesis Guide

This appendix contains the site visit interview protocols and the synthesis guide in their original form.

MILITARY PROTOCOL

General Questions Related to Emergency Management and Planning for Major Disasters

1. What is your current position, and in what office or organization are you based?
   - What are your major responsibilities with respect to disaster planning and emergency management?
   - Over the course of your career, how long have you been involved with aspects of major disaster and emergency planning?

2. How does your office contribute to [installation name's] overall role in emergency management and responding to major disasters?

3. For the purposes of major disaster planning and emergency management, how does [installation name] define the term "community?"
   - [Prompt if needed: For the purposes of major disaster planning, who does (organization name from Q1) consider as the community?]
   - [Probe if needed:] What groups of people or organizations fall within [organization name's] definition of community?
   - [Probe if needed:] Do geographic boundaries factor into [organization name's] definition of community?

4. Broadly speaking, how would [installation name], civilian organizations, and the VA collaborate or otherwise interact in the event of a major disaster or emergency in the surrounding community (outside the gates of the installation)?
   - How would that change, if at all, if military assets were involved in the event?
   - To what extent does [installation name] look to the Guard or the Reserve for local assistance in responding to disasters and emergencies in the community?
   - To what extent do other [Service] elements play a role in responding to an emergency in [installation name's] local community? For example, do regional or other higher-level organizations aid [installation name] in response planning or even participate directly in community-level exercises?

5. Has your community experienced a major disaster (even small-scale events that could have a potential catastrophic consequence—e.g., a hurricane or a fire)? [If yes:] Please describe. [If no, skip to question 7.]

6. How did [installation name] assess risk for this event prior to the experience, if at all?

- Were there plans in place for the capabilities that would be needed for response in this type of event? [If yes:] Please describe how that worked.

Plans and Exercise

7. How is [installation name] currently planning for major disasters? Do [installation name] personnel plan for specific threats (which one/s?), or do they take more of an all-hazards approach?

- To what extent are [installation name] plans in writing? [If needed, follow up on RAND's advance request for copy.]
- We've learned that disaster response planning at [installation name] involves different personnel working on issues related to anti-terrorism, public health emergencies, military support for civilian authorities, and the like. To what extent do personnel tasked with these different responsibilities plan together?
- Have personnel from [installation name] developed plans or engaged in other planning-related activities with either local civilian organizations or the VA?
- [If multiple military installations in the area:] What about with nearby installations? Has [installation name] engaged in planning-related activities with them?

8. [If exercises not discussed in Q7:] How do exercises and training tie in to the planning process at [installation name] and with the community more broadly?

- How often does [installation name] conduct emergency response exercises?
- We already discussed their interaction with respect to planning, but to what extent do [installation name] personnel who work on different issues, such as public health, anti-terrorism, and military support for civilian authorities exercise together?
- Have personnel from [installation name] participated in exercises or training with local civilian organizations or the VA? [If no:] Why not? Has [installation name] fielded requests to participate in such activities?

- [If multiple military installations in the area:] What about with nearby installations? Has [installation name] participated in exercises and other training with them?

9. Are there any legal or governance issues that affect military-civilian or military-military interactions in your community? [If yes:] How so?

10. Similarly, do any interoperability issues affect military-civilian or military-military interactions in your community? [If yes:] How so?

11. What DoD or [Service] guidance is [installation name] currently following in its planning for major disasters? This guidance may include publications related to anti-terrorism, public health emergencies, military support of civil authorities, or general emergency management.

- [Always ask:] Are there other military sources of disaster preparedness guidance, such as regional or major commands? [If yes:] Please describe.

12. What non-DoD guidance or frameworks is [installation name] currently using in its planning for major disasters?

- [Prompt if needed: Examples of policies and mandates may include guidelines from the Cities Readiness Initiative, the Urban Areas Securities Initiative, Bioterrorism funding, the National Response Framework, National Planning Scenarios, and the Target Capabilities List.]
- [If yes:] How helpful is it/are they?
- [If no:] Why not?

13. How useful is the guidance provided to [installation name] from these different sources?

- Are there instances in which guidance from different sources conflicts? [If yes:] How so?
- What is lacking from DoD and [Service] guidance? In other words, what aspects of planning need to be addressed but have not been, and which are in need of further clarification?

Risk to Your Community

We're interested in how [installation name] determines or otherwise considers your community's risk for major disasters and how that may influence preparedness plans. Risk is often thought of in terms of threats, vulnerabilities, and consequences, and planning is directed to capabilities. We have some questions about these topics today. Let's start with threats first.

Threats

14. How does [installation name] determine threats to the community?
    - Are different types of threats assessed by different personnel? [If yes:] How so?
    - Do personnel at [installation name] use National Planning Scenarios or some other guidance?
    - [Prompt if needed: The scenarios include pandemic influenza, natural disasters like hurricanes and earthquakes, and terrorist attacks using different means such as biological, chemical, nuclear, and explosives.]

Vulnerabilities and Consequences

15. How does [installation name] determine your community's vulnerabilities?

16. How does [installation name] identify which assets are most essential to the community—in other words, those for which the consequences would be most detrimental?
    - Where did [installation name] obtain the information that guided its decisions about key assets?

Capabilities

17. What are the major/primary capabilities [installation name] can contribute or offer in support of responding to major disasters within your community? We're particularly interested in response-related capabilities included on the Target Capabilities List, such as on-site incident management, public safety and security, and mass care.
    - How would [installation name's] capabilities to respond differ, if at all, across the various events outlined in the National Planning Scenarios?

- [Prompt if needed: The scenarios include pandemic influenza, natural disasters like hurricanes and earthquakes, and terrorist attacks using different means such as biological, chemical, nuclear, and explosives.]

18. Which of these capabilities are critical for [installation name] to provide the community in an emergency or major disaster event?
    - Which of these capabilities, if any, can only [installation name] provide?

19. What might jeopardize the availability of [installation name's] capabilities?
    - Do they change based on constraints like funding, OPTEMPO, other mission requirements, and the like?

Feedback on Decision Support Tool

As I mentioned earlier, the primary outcome of our study is a decision support tool. The decision support tool itself would be developed in a follow-on effort, and ultimately both military and civilian organization in your community—as well as those in other communities—could use this tool to assess risk and to identify needed capabilities. The tool would require input of data on local assets and capabilities, threats, and vulnerabilities.

20. To the best of your knowledge does [installation name] currently use or has it previously used a formal planning tool to plan for major disasters? This may include, but is not limited to, computer-based models or software.

21. How useful would a new decision support tool be to aid [installation name] personnel in their response planning?
    - Why/why not?

22. What features would you like to see in such a decision support tool?
    - What types of functionality would be particularly beneficial or important to [installation name]?
    - [If question 21 had an affirmative response, ask:] How could a new decision support tool improve or otherwise build upon previous or current planning tools used at [installation name]?

23. What data do you think should be included in this type of decision support tool?

- How much of these data are currently available? We're interested in data available locally as well as any state or national level data sources of which you are aware.
- Which of these data could [installation name] personnel input in this type of tool?

24. How challenging would it be to gather this type of data, particularly for [installation name] and others in the community?

25. Does [installation name] have resources available to maintain a tool that requires the input of local data?

26. What could limit the use of a decision support tool for local military and civilian planners and responders?
  - For example, could this tool be developed and maintained at the unclassified level?
  - Would interoperability issues pose a problem?
  - Are there ways these challenges could be avoided or overcome?

27. Those are all the questions that we have for you today. Is there anything you would like to add regarding [installation name's] emergency management efforts in particular or those of your community more generally? Anything that we did not ask but should have?

CIVILIAN PROTOCOL

General Questions Related to Emergency Management and Planning for Major Disasters

1. What is your current position, and in what office or organization are you based?
  - What are your major responsibilities with respect to disaster planning and emergency management?
  - Over the course of your career, how long have you been involved with aspects of major disaster and emergency planning?

2. What is [your organization's] role emergency management and responding to major disasters?

3. For the purposes of major disaster planning and emergency management, how does [organization name] define the term "community?"

- [Probe if needed:] What groups of people or organizations fall within [organization name's] definition of community?
- [Probe if needed:] Do geographic boundaries factor into [organization name's] definition of community?

4. Broadly speaking, what is the role of the military, namely [local installation name(s)] and the VA should a major disaster occur in the surrounding community?

- To what extent does [organization name] look to the Guard or the Reserve for local assistance in responding to disasters and emergencies in the community?

5. Has your community experienced a major disaster (even small-scale events that could have a potential catastrophic consequence—e.g., a hurricane or a fire)? [If yes:] Please describe. [If no, skip to question 7.]

6. How did [organization name] assess risk for this event prior to the experience, if at all?

- Were there plans in place for the capabilities that would be needed for response in this type of event? [If yes:] Please describe how that worked.

Plans and Exercises

7. How is [organization name] currently planning for major disasters? Do [organization name] staff plan for specific threats (which one/s?), or do they take more of an all-hazards approach?

- To what extent are [organization name] plans in writing? [If needed, follow up on RAND's advance request for copy.]
- Have staff from [organization name] developed plans or engaged in other planning-related activities, with local installations or the VA?

8. [If exercises not discussed in Q7:] How do exercises and training tie in to the planning process at [your organization] and with the community more broadly?

- How often does [organization name] conduct emergency response exercises?
- Have staff from [organization name] participated in exercises or training with local [installation names] or the VA? [If no:] Why not?

9. Are there any legal or governance issues that affect military-civilian or military-military interactions in your community? [If yes:] How so?

10. Similarly, do any interoperability issues affect military-civilian or military-military interactions in your community? [If yes:] How so?

11. What guidance or frameworks is [organization name] currently using in its planning for major disasters?

- [Prompt if needed: Examples of policies and mandates may include guidelines from the Cities Readiness Initiative, the Urban Areas Securities Initiative, Bioterrorism funding, the National Response Framework, National Planning Scenarios, and the Target Capabilities List.]
- [If yes:] How helpful is it/are they?
- [If no:] Why not?
- Are there instances in which guidance from different sources conflicts? [If yes:] How so?

Risk to Your Community

We're interested in how [organization name] determines or otherwise considers your community's risk for major disasters and how that may influence preparedness plans. Risk is often thought of in terms of threats, vulnerabilities, and consequences, and planning is directed to capabilities. We have some questions about these topics today. Let's start with threats first.

Threats

12. How does [organization name] determine threats to the community?

- Are different types of threats assessed by different staff? [If yes:] How so?
- Do staff at [organization name] use National Planning Scenarios or some other guidance?
- [Prompt if needed: The scenarios include pandemic influenza, natural disasters like hurricanes and earthquakes, and terrorist attacks using different means such as biological, chemical, nuclear, and explosives.]

Vulnerabilities and Consequences

13. How does [organization name] determine your community's vulnerabilities?

14. How does [organization name] identify which assets are most essential to the community—in other words, those for which the consequences would be most detrimental?

- Where did [organization name] obtain the information that guided its decisions about key assets?

Capabilities

15. What are the major/primary capabilities [organization name] can contribute or offer in support of responding to major disasters within your community? We're particularly interested in response-related capabilities included on the Target Capabilities List, such as on-site incident management, public safety and security, and mass care.

- How would [organization name's] capabilities to respond differ, if at all, across the various events outlined in the National Planning Scenarios?
- Prompt if needed: The scenarios include pandemic influenza, natural disasters like hurricanes and earthquakes, and terrorist attacks using different means such as biological, chemical, nuclear, and explosives.]

16. Which of these capabilities are critical for [organization name] to provide the community in an emergency or major disaster event?

- Which of these capabilities, if any, can only [organization name] provide?

17. What might jeopardize the availability of [organization name's] capabilities?

- Do they change based on constraints like funding, other organization requirements, and the like?

Feedback on Decision Support Tool

As I mentioned earlier, the primary outcome of our study is a decision support tool. The decision support tool itself would be developed in a follow-on effort, and ultimately both military and civilian organization in your community—as well as those in other communities—could use this tool to assess risk and to identify needed capabilities. The tool would require input of data on local assets and capabilities, threats, and vulnerabilities.

18. To the best of your knowledge does [organization name] currently use or has it previously used a formal planning tool to plan for major disasters? This may include, but is not limited to, computer-based models or software.

19. How useful would a new decision support tool be to aid [organization name] personnel in their response planning?
    - Why/why not?

20. What features would you like to see in such a decision support tool?
    - What types of functionality would be particularly beneficial or important to [organization name]?
    - [If question 19 had an affirmative response, ask:] How could a new decision support tool improve or otherwise build upon previous or current planning tools used at [organization name]?

21. What data do you think should be included in this type of decision support tool?
    - How much of these data are currently available? We're interested in data available locally as well as any state or national level data sources of which you are aware.
    - Which of these data could [organization name] staff input in this type of tool?

22. How challenging would it be to gather this type of data, particularly for [organization name] and others in the community?

23. Does [organization name] have resources available to maintain a tool that requires the input of local data?

24. What could limit the use of a decision support tool for local military and civilian planners and responders?
    - For example, could this tool be developed and maintained at the unclassified level?
    - Would interoperability issues pose a problem?
    - Are there ways these challenges could be avoided or overcome?

25. Those are all the questions that we have for you today. Is there anything you would like to add regarding [organization name's] emergency management efforts in particular or those of your community more generally? Anything that we did not ask but should have?

CIVILIAN PROTOCOL: EXECUTIVE BRANCH OVERSIGHT AND DELEGATION

General Questions Related to Emergency Management and Planning for Major Disasters

These questions focus on oversight and delegation (e.g., to whom/what departments does [organization] delegate to or provide oversight for in terms of planning, response, mitigation and threat assessment, vulnerability assessment and risk assessment?).

1. What is your role and what are your organization's major responsibilities with respect to disaster planning and emergency management?
   - What is the nature of the relationship between your office and the operators/administrators such as police, fire, [and] public health in terms of lines of authority? (e.g., City is tasked as the "Director of EM"; Fire Department is tasked as "Administrator" of EM)
   - Is there a formal document that outlines these relationships and may we have a copy?
   - What is the nature of your role in the following: Identification of critical infrastructure/critical assets and/or vulnerabilities in your community?
   - Where did [organization name] obtain the information that guided its decisions about key assets?
   - Threat assessment in your community?
   - Risk assessment for specific events in your community?
   - What is the nature of your relationship to the VA and local military installations? Or does that interaction occur through the administrator?
   - What is the nature of your relationship to the state (Governor's office, legislature) regarding emergency management and homeland security?
   - What is the nature of your relationship to DHS, FEMA, CDC or other federal homeland security–related organizations? Do you have contact with them directly or do those interactions occur through the designee/administrator?

2. What guidance documents (e.g., laws, policy directives at the state, federal level) do you see as establishing your requirements for Emergency Management and Homeland Security?

Lines of Communication/Network

If not covered in questions on oversight and delegation, may want to follow up with question about "regular" contact and communication.

3. With whom do you directly communicate most often with respect to EM? What is the nature of that communication?
   - Assuming you've assigned primary responsibilities for EM Administration to Fire with the support of Police and Public Works, with which other organizations/offices/agencies do you have regular contact?

Budgeting and Reimbursement

These questions focus on budgeting, reimbursement and grants related to emergency management and homeland security.

4. How does the decision maker work with local emergency management response organizations in the initial budgeting process?
   - How does it work if a responding organization needs to establish a new program (e.g., public awareness) or they need new equipment (chemical detection) and hence a larger budget in a given fiscal year?

5. How do reimbursements for disasters work?
   - Is there a staff person who coordinates requests for state and local disaster reimbursement funding from the City?

6. What happens if the responders spend more responding to a specific incident than the City has budgeted?

7. Who is responsible for writing grant applications to DHS and similar organizations?

Decision Support Framework or Other Tools

8. Are you familiar with, or does your organization use, a formal planning tool to plan for major disasters? (This may include, but is not limited to, computer-based models or software.)
   - [If yes:] How do you use the tool, either alone, or in conjunction with your operational staff? How does it get funded and disseminated?

9. Are there any tools that you have seen, or that you could envision, that might enable your office to function more efficiently or effectively with respect to your specific EM responsibilities? (Examples might include anything from better guidance to specific tools.)

10. Is there anyone we haven't captured in our interviews thus far that you believe is critical to understanding the breadth of responsibilities in EM/HLS (e.g., Public works? Sheriff's Department)?

SITE-VISIT SYNTHESIS GUIDE

The purpose of this guide is to provide the backbone for a systematic examination of the evidence collected from each of the five sites, first within each site and then across sites. For each substantive heading or topic, team members should independently evaluate the data collected—primarily from interviews for most topics—and craft roughly five bullets (sometimes more, sometimes less) that capture the most-compelling themes for that topic. The goal is to identify themes across data sources. Special attention should be paid to identifying findings with a high level of agreement (either overall or within a certain group, such as military or those functionally similar) as well as to those with striking contrasts between stakeholder groups. The absence of data or a consistent lack of a response to a question may also be worth noting. Less emphasis should be placed on findings that emerge from one only data source; such findings are likely more tentative and possibly best presented as useful orienting background or interesting/unique features, where they will appear more as analytic "food for thought."

While not all bullets need annotation in terms of a source(s), it may be helpful to note the proportion of interviews in which a finding was present or to cite a specific interview in the event of a controversial or unique point that others may wish to look up directly. This also helps to ensure that findings are selected based on evidence quality rather than salience or another bias. It may also be useful to provide a breakdown by military/community or function to the extent that is relevant. Military-specific or community-specific findings definitely should be noted as such.

After bullets are developed independently, they will be integrated into one document and ultimately consolidated into a tighter set of bullets that reflect the overall details of the site and the consensus of the site-visit team members. The bullets, in turn, will serve as the foundation for the cross-case analysis as well as for the development of a site-specific report appendix.

Data Sources

(A short list of the materials that contributed to this summary, including a breakdown of interviews by number and/or type, details regarding network survey responses, and any relevant documents. Any gaps in data collection should also be noted here.)

Useful Orienting Background

(Interesting, potentially relevant background that may help set the stage for subsequent findings; may include information regarding level of expertise/training of interviewees, experience with major disasters in the past, or overarching observations by interviewees. If we don't include the site selection criteria as a table [to present stats like population, funding, FEMA region], incorporate that here.)

Definitions of Community

(Includes findings pertaining to uniformity—or variation—of community definition, especially given different types of disasters, the mil/civ distinction, and functional diversity of interviewees.)

Disaster Preparation: Plans and Exercises

(Includes findings about the current planning process such as event specificity; exercise nature and frequency; interplay between plans and exercises; the nature of interaction between civilian stakeholders, between military stakeholders, and/or across mil/civ boundaries; involvement of other local elements such as VA, Guard, or Reserve.)

Disaster Preparation and Response Facilitators

(Covers resources that foster planning or exercises, to include DoD and non-DoD guidance, extant decision support tools, and other facilitators, both tangible [e.g., grants, MOUs] and intangible [e.g., social connections]. While envisioned more as obstacles, legal/governance and interoperability issues may also be discussed here.)

Disaster Preparation and Response Obstacles

(Covers hindrances to planning or exercises, to include legal/governance and interoperability issues along with other impediments, both tangible [e.g., financial constraints] and intangible [e.g., resistance to change]. While envisioned more as facilitators, guidance and extant decision support tools may also be discussed here. Also includes strategies, both potential and applied, used to overcome these obstacles.)

Threat Assessment

(Includes findings related to who conducts this assessment for different types of events; how it is accomplished; reliance on outside guidance; frequency with which assessments are re-visited; the extent to which threat assessment data are shared; and potential input from state or federal stakeholders.)

Vulnerabilities Assessment

(Includes findings related to who conducts this assessment; how it is accomplished; information sources for assets; frequency with which assessments are re-visited; the extent to which vulnerability assessment data are shared; and potential input from state or federal stakeholders.)

Capabilities

(Features assessments of the primary capabilities offered by major stakeholders, both military and civilian; factors that may influence the availability of such capabilities [e.g., event type, OPTEMPO]; and perceptions of critical or unique capabilities.)

Decision Support Tool: Potential Applications

(Covers perceived usefulness of a new design support tool [both pro and con], especially vis-à-vis other aids; perceptions of beneficial features; ideas about how/when such a tool might be used; and opportunities to improve or otherwise build upon features of extant tools.)

Decision Support Tool: Implementation-Related Issues

(Covers factors that could help or hinder the development and subsequent implementation of a decision support tool, such as current availability of local data, interoperability issues, data sensitivity and classification, perceptions of maintenance required, and data safeguarding. Also includes recommendations for successful implementation and cautionary advice.)

Site Visit Interesting or Unique Features

(Captures unique or noteworthy aspects of this site; implications of this case for cross-case analysis and for the overall study; and interesting observations that may be thought-provoking yet not prevalent among data sources.)

APPENDIX E

Site-Specific Interview Summaries

As discussed in Chapter Four, site visits were conducted for five communities to inform the development of our decision support tool framework. RAND researchers spent one to three days at each location to conduct in-person interviews, primarily at the local military installation(s), and additional interviews were conducted via telephone after visiting each location to ensure that both military and civilian perspectives on local disaster preparedness were captured. Our site visits were completed in the following order, with the timing of the actual location visit in parentheses:

- San Antonio Metropolitan Area, Texas (April 2008)
- Norfolk and Virginia Beach Metropolitan Area, Virginia (May 2008)
- City of Columbus and Muscogee County, Georgia (July 2008)
- City of Tacoma and Pierce County, Washington (September 2008)
- City of Las Vegas and Clark County, Nevada (September 2008).

While Chapter Four features the results of our cross-site analysis, including notable similarities and differences across the five communities, this appendix features the more detailed results of the within-site analysis for each location that facilitated the cross-case analysis. The within-site analysis was completed independently by two RAND researchers, typically one who visited the location and one who did not. They each used the site visit synthesis guide provided in Appendix D in their analysis of expert-interview data. After independently drafting bullets to summarize the interview data from both military personnel and civilian-agency employees at a specific site, the two researchers developed a consolidated version of their summaries that reconciled any differences and integrated redundancies. The results were vetted by a third RAND researcher, who had participated in the site visit.

The sections of this appendix are sequenced in order of the site visits, and each section is organized roughly according to the synthesis guide provided in Appendix D. Note that the information in these site visit summaries is based on the perceptions of expert-interview participants at the time of our research. While efforts have been made to resolve inconsistencies across interviews and to note updates to civilian guidance and military doctrine, we did not routinely seek independent verification of the information provided during expert interviews. In addition, we acknowledge that local disaster preparedness efforts and the actual events experienced likely have changed since we collected these interview data. Nevertheless, these summaries offer an informative, multifaceted perspective of five communities' experience with disasters, both perceived and actual, and disaster planning efforts as of 2008.

San Antonio Metropolitan Area, Texas (April 2008)

Background

Texas is a large state, both by land area and by population. Consequently, preparedness planners face some special challenges. San Antonio emergency management staff are responsible for 22 counties spreading south to the U.S.-Mexico border, covering tens of thousands of square miles. Houston is 200 miles away, and San Antonio is distant from other major urban areas. San Antonio has the only HAZMAT team for hundreds of miles.

There are many federal assets in the region, including two Air Force bases (Randolph and Lackland), an Air Force Major Command (the Air Education and Training Command, or AETC), an Army post (Fort Sam Houston), and a major military hospital (Brooke Army Medical Center, or BAMC). Fort Sam Houston is located in the middle of San Antonio; Randolph AFB is technically not within the city limits. San Antonio is also home to many resources from the VA, the Navy, and the former Kelly AFB (now Kelly Annex under Lackland AFB), which supports Air National Guard and Air Force Reserve missions as well as the city of San Antonio.

Together, Fort Sam Houston and Lackland and Randolph AFBs cover almost 9,000 acres. Lackland AFB has the largest population, at nearly 19,000 active duty personnel, while Randolph AFB has approximately 3,600 and Fort Sam Houston about 8,000. Lackland AFB is home to the 37th Training Wing, which is the largest training wing in the U.S. Air Force, providing basic military, professional, and technical skills to the Air Force, other services, government agencies, and allies. Randolph AFB is home to the Air Force's AETC, which also provides basic military training as well as initial and advanced technical training, flight training, and graduate-level military training. Fort Sam Houston is the command headquarters for the 5th U.S. Army and U.S. Army South and the home of BAMC.

Organizations in the San Antonio area are recipients of several HHS grants, including the CRI, Laboratory Bioterrorism Initiative, and the CDC cooperative agreement related to Public Health Emergency Preparedness. San Antonio is considered a UASI city, and local agencies also receive funds from DHS and through the MMRS program.

The respondents' disaster experience includes hurricanes, heavy flooding (especially 1998 and 2000), a chlorine tanker spill in 2003–2004 with three fatalities, and a water-supply situation at Randolph AFB in 1999–2000, during which it lacked a potable water source and received support from the city. Reported potential threats include flooding, hurricanes, terrorism (especially with military and railroad resources nearby), high temperatures, drought, foodborne illness, and infectious disease, since the Alamo attracts many tourists. Flooding is the greatest threat to organizations in the San Antonio area; however, interview participants recognized that having a large presence of military personnel and assets located in San Antonio/Bexar County makes terrorism a potential concern as well.

Interviewees from the area have a lot of experience with hurricanes; thus, they place primary emphasis on plans and exercises related to hurricane response. Although the area is inland and not usually threatened by major storms, San Antonio–area residents and organizations take in many evacuees. Lackland AFB plays a pivotal part in hurricane support for the community and is part of the NDMS supported by HHS. However, Randolph AFB can accommodate only DoD personnel and dependents, since its housing capacity is limited.

During Hurricane Katrina, the San Antonio area was a major relocation and shelter site. Lackland AFB staff communicated immediately with DoD officials and started operating around the clock. It was a staging area for evacuees, provided its runway for transport of persons and equipment, and supplied manpower for Kelly Annex, the former Air Force base that housed evacuees. Randolph AFB was involved in policing activities, offered support from its helicopter squadron, and sent more than 200 personnel to support civilian emergency management. Randolph AFB also provided temporary housing for families and personnel from Keesler AFB in Mississippi.

Aside from hurricane support, organizations in the San Antonio area plan and exercise for regularly occurring high-profile events. For example, the annual air show draws 250,000 people to the area. Local fire and HAZMAT teams are available and have different command posts set up around the base. San Antonio police and security forces have crowd control and lock-down protocol in place in the event of an emergency. The NCAA Final Four tournament was also held San Antonio in 2008, and public safety and health officials prepared for an emergency knowing that most visitors would not have vehicles.

Interview Profiles

RAND's research team conducted 14 interviews with professionals from Randolph AFB, Lackland AFB, Fort Sam Houston, the city of San Antonio, and Bexar County. At Randolph AFB, participants represented such functions as emergency management, security, antiterrorism, CBRNE, an exercise-evaluation team, plans and programs, public health, and bioenvironmental engineering. Public health, bioenvironmental-engineering, antiterrorism, medical-readiness, planning, exercise-evaluation team, and civil-engineering representatives were interviewed at Lackland. The small number of interviews at Fort Sam Houston included staff responsible for planning, operations, public health, medical, antiterrorism, public safety, and public affairs. Finally, the RAND team interviewed San Antonio civilian authorities responsible for emergency management, fire, MMRS, public health, and public safety, and a representative from the VA. The RAND team was unable to conduct a decisionmaking-oriented interview with an executive-branch representative.

Definitions of Community

Civilian interviewees consistently defined the community as extending beyond the city of San Antonio and its citizens. This broad view probably stems primarily from San Antonio's location. Major urban areas are distant, and surrounding counties do not have many resources. As interview participants noted, San Antonio's designation as a "centroid" for much of the hurricane response in Texas brings several state players to the table on a regular basis, and these interactions bleed into other aspects of a public health and terrorism emergency. The city of San Antonio also specifically tries to consider the military-related personnel and families who may be off base during an emergency.

Both military and civilian public health and medical personnel tended to cast the community in wide terms. BAMC staff felt accountable to all of Trauma Region P, which includes 53 hospitals in 80 counties. BAMC interviewees also referred to South Texas Regional Area Community Health (STRACH), a quasi-formal organization in the greater San Antonio area that deals with emergency issues. BAMC contributes a health physicist and an epidemiological nurse to work with the collaboration, which holds meetings and coordinates disaster supplies. San Antonio Metropolitan Health District personnel commented that the community not only extends past city or even county lines, but includes tourists as well. Public health participants further stated that the SNS includes 1.5 million people in Bexar County, as well as up to 35,000 people who may be visiting for business.

On the military side, security personnel had the narrowest vision of what the community entails, focusing first and foremost in protecting people and assets within the base. Their role may extend outside the installation for an immediate response or as first responders in a situation involving a DoD asset (e.g., a military plane crash) and the establishment of a National Defense Area. The antiterrorism aspect of operations included a larger community of intelligence sharing and threat assessment with other stakeholders, such as the Air Force OSI, the FBI, and local law enforcement.

The VA has regional emergency response management and planning responsibilities for a significant portion of South Texas. Its area of responsibility runs from San Antonio south through Corpus Christi to Brownsville, and west to Laredo. The VA serves San Antonio city, county, and the military region, along with the city of Kerrville and Kerr County. It also has outpatient clinics in Laredo, McAllen, Harlingen, Corpus Christi, Victoria, and San Antonio. South Texas Veterans Health Care System is one of three systems in VISN 17, centered in Dallas.

Community Risk Assessment

Threat Assessment. On the military side and possibly the civilian side as well, threat assessments determine criticality or consequence assessment, which leads to vulnerability assessment and the subsequent overall risk assessment. On both Air Force and Army installations, threat assessment varies depending on the nature of the threat, and different personnel or offices evaluate different kinds of threat. The installation ATO conducts an annual terrorist-focused threat assessment, with input from military entities, such as military intelligence, Air Force OSI, Army Criminal Investigation Command (CID), and installation-level security forces, as well as civilian stakeholders, such as the local police and the FBI. Fort Sam Houston interview participants also specifically mentioned the JTTF as a participant in the threat assessment process.

To the extent that other non-terrorist threats are assessed, they are handled by specific functional offices (e.g., public health considers the pandemic influenza threat, CBRNE assesses threats related to toxic industrial chemicals and materials transported via rail). The specific functions involved vary by service. For example, the Army does not have a direct equivalent to the Air Force bioenvironmental engineers. It was not clear from the interviews whether each of the San Antonio–area military installations has developed a composite installation threat assessment that addresses all types of threats. Although civilian actors may contribute to a military threat assessment process, there is no promise that the results of the military assessment will be shared with civilian authorities, and much of the assessment may only be available at a classified level (secret or higher). Information sharing relates more to intelligence about different types of threats rather than to the assessment itself.

San Antonio's emergency management personnel claimed that the office did some of the threat assessments recommended by DHS through the Buffer Zone Protection Plan (BZPP) grants.[1] Emergency managers have to adhere to DHS tier 1 and 2 critical infrastructure lists. Public health officials carry out risk assessment in conjunction with the city's Office of Emergency Management, the Texas Department of State Health Services, and the Texas Division

[1] The BZPP provides funding for the planning, equipment, and management of protective actions, with the objective of protecting, securing, and reducing the vulnerabilities of identified critical infrastructure and key resource sites (authorized by DHS).

of Emergency Management. The city's Office of Emergency Management identifies threats within the community and forwards them to the public health department, whose staff then develop a response plan in collaboration with emergency management personnel. Public health personnel receive results of emergency management assessments periodically, but there is no formal agreement that they should receive them. Generally, it is during the EOC staff meeting that a threat is identified. In that case, a determination is made whether the threat is already covered within the current all-hazards plan or whether a new, scenario-specific plan needs to be developed, as was the case for the NCAA Final Four tournament event.

The VA essentially drafts a threat list based on committee opinion. The VA emergency management committee is made up of individuals representing functions identified in the VA guidebook, and they consider both threats in the local area as well as in surrounding locations. VA employees consider the likelihood of an event occurring directly or in the vicinity under each category listed on their hazard vulnerability assessment tool and rate probabilities from low to high on a three-point scale for that category. Staff also discuss potential events (e.g., terrorism, major chemical event on nearby highway) and what might affect the VA. The San Antonio VA does the initial threat assessment, then sends the assessment to its Emergency Management Committee for review; changes may be made and alternatives for discussion identified.

Vulnerability Assessment. According to Air Force personnel, military vulnerability assessments are performed by the installation threat working group (TWG) after threats and criticalities are identified. Installations do not address community-level issues in detail but complete a comprehensive vulnerability-assessment plan (CVAP) for the installation. There is a classified DoD database to document vulnerabilities. Every installation performs an annual vulnerability assessment. The Defense Threat Reduction Agency (DTRA) is DoD's lead agency for vulnerability assessment. DTRA teams perform JSIVAs. Each installation is required to have a DTRA JSIVA every three years.

Lackland AFB personnel mentioned Guardian Lite as a resource for vulnerability assessment and noted that they constantly interface with Air Force OSI, the San Antonio Police Department, and additional civilian agencies as appropriate to assess any vulnerability. Randolph AFB (and the major command, or MAJCOM, under which it falls) uses a defined risk assessment approach to help prioritize mitigation recommendations for the installation commander to advocate for funding. The process follows a standard methodology: threat assessment, criticality assessment, vulnerability assessment, then risk assessment. AETC uses a tool called Force Pro for this process. Interviewees also told the RAND study team about a person at the Air Force Research Laboratory at Eglin AFB who is developing a multidisciplinary risk assessment tool that they indicated could be fielded as early as late 2009.

As for community vulnerability assessments, civilians pull data from DHS tier 1 and 2 critical infrastructure lists. Public health personnel also conduct vulnerability assessments and all the critical infrastructure facility assessments. Civilian public health and fire departments have a GIS unit that assesses hazards; this analysis is shared with the City of Houston. They use a tool called Digital Sandbox, mapping where assets and clusters of greatest risk are located. The VA conducts its own type of vulnerability assessment, as part of its hazard vulnerability assessment process, described earlier.

Local Planning and Exercises

Overall, planning in the San Antonio region appears to be an annual process, and exercises occur year round in various formats (e.g., tabletop, functional, comprehensive) with various actors involved. Both the military and civilian sides have exercise requirements to be met, and there is a general sense that exercises are designed to test and vet plans.

The State of Texas has a template for basic emergency response plans and 22 functional annexes. San Antonio is using this to develop standard operating guidance, with the assistance of the Texas Engineering Extension Service (TEEX) of Texas A&M University. The state template also features an annual review and recertification every five years. Furthermore, San Antonio emergency management staff conduct training and exercises for the state and, at the time of our interviews, were compiling an AAR for Hurricane Dolly. All 22 planning annexes go through a coordination process at the weekly emergency planning meetings, which include government officials, the Red Cross, the private sectors, and universities. Other organizations, such as Texas A&M and National Emergency Response and Rescue Training Center, provide planning support. When hurricane season approaches, this is the primary preparedness focus; otherwise, San Antonio has an all-hazards planning process. Although all exercises must be coordinated through TEEX, regardless of grant source, the city has the ability to steer them and use "injects" to highlight particular areas of interest. It often will inject deficiencies identified in previous exercises into the scenario for a subsequent exercise so that it can explicitly exercise ways to overcome such deficiencies.

Both Lackland and Randolph AFBs have comprehensive all-hazards plans for emergency management, CEMP 10-2,[2] which contains a number of functional and event-specific annexes. The Army base does not have a similarly global plan; instead, planning is conducted in a central organization at the installation, and all the major tenant units feed their plans into that office. However, the end result seems to be plans that are both all-hazards and event specific. At Lackland AFB, each functional organization has its own checklist based on the aforementioned 10-2. The military first responders (e.g., fire, medical, HAZMAT) meet quarterly to bring together all information to report to the commander. Medical personnel at the installations also develop their own robust plans, such as the MCRPs at the Air Force bases.

---

[2] According to AFI 10-2501,

> The installation CEMP 10-2 provides comprehensive guidance for emergency response to physical threats resulting from major accidents, natural disasters, conventional attacks, terrorist attack, and CBRN attacks. As such it is intended to be a separate installation plan and will not be combined with other plans until HQ USAF [headquarters U.S. Air Force] develops and fields a template and provides implementation guidance. All installations must develop a CEMP 10-2 using the AF template to address the physical threats to their base. . . . The CEMP 10-2 should be coordinated with . . . other installation plans such as the AT Plan 31-101, Base Defense Plan, MCRP, ESP and Installation Deployment Plan. The CEMP 10-2 must be coordinated through all tasked agencies and should be coordinated with all units/agencies on the installation. Any conflicts with other plans must be resolved before publication. Readiness and Emergency Management Flights will provide an information copy of the CEMP 10-2, unless it is classified, to local civilian agencies as part of their total coordination effort.

In December 2008, eight months after our site visit to the San Antonio area, Army Regulation 525-27: Military Operations Army Emergency Management Program, was published. That likely has affected the emergency management planning process employed by Fort Sam Houston personnel and the output produced.

Key Actors and Coordination in Local Planning

Civilian interviewees suggested that the military has become more involved in civilian emergency management in recent years. At the time of our research, military representatives were part of the community EOC and at the table for civilian exercises. However, the military is typically not assigned tasks in community plans, in part because it is not clear what capabilities the military will be able to contribute, given fluctuations in the military's local capabilities (due to OPTEMPO), its inward focus, and its emphasis on counterterrorism. Although the military does not share its plans with civilians, local civilian officials reported that the military has signed off on some of the relevant annexes in the civilian emergency preparedness plan and that civilian and military authorities have established MAAs. Fire officials also have robust military and civil interaction, though their purpose is potentially more for response than for planning. In many incidents, a representative from civilian fire will serve as the incident commander.

Interviews indirectly suggested that the medical and public health sectors have engaged in more collaboration and coordination, both military-military and military-civilian, than other functional areas. This was also explicitly noted at Fort Sam Houston and by San Antonio Police Department personnel. Groups that focus on medical and public health planning and response include the San Antonio Emergency Medical System and Hospital Disaster Group, the NDMS, and the Regional Medical Operations Center. Community medical facilities, military medical facilities, and the VA are all very involved in these types of coordination networks. Furthermore, public health employees work specifically with military personnel to coordinate their SNS plans. The military is not audited by the CDC, but it is part of the SNS process in terms of enumerating population and mass prophylaxis distribution methods.

Fort Sam Houston staff appear to collaborate more with their civilian emergency management counterparts than do staff from the local Air Force bases. While the Army is not interacting with civilians to the point of engaging in joint planning, Fort Sam Houston does have a liaison officer for the local community EOC and a seat at the table at the Alamo Area Council of Governments. The Alamo Area Council of Governments included major external players in its installation exercise strategy, including liaison officers from local fire and police agencies. Fort Sam Houston also contributed to community exercises, such as a chemical drill and an airplane crash scenario at the San Antonio airport. In addition, civilian public health officials noted that the city and county have extensive MAAs with Fort Sam Houston.

Conversely, Air Force interviews revealed that Air Force headquarters is not mandating or even promoting joint combined installation-community planning; some Air Force interviewees noted that interactions with civilian counterparts were infrequent. It also appears that the Air Force has less involvement in the civilian emergency preparedness infrastructure, such as the city EOC and the Alamo Area Council of Governments. Nonetheless, one cannot conclude that the Air Force is not supporting civilian agencies in disaster preparedness. The Air Force—Lackland AFB in particular—does offer capabilities to assist the civilian community and participates in collaborations at a functional level. Further, there is a sense, at least among some civilian stakeholders, that the military has become more involved in civilian disaster preparedness efforts since Hurricane Katrina. As mentioned earlier, county officials work with Lackland AFB because of its active runway and designation as an NDMS evacuation hub. Civilian emergency preparedness employees also engage Lackland AFB in disaster planning because it is a consolidation point for evacuating Southeast Texas, it has consistent points of contact in emergencies, and it would be critical actor in air transport and traffic control during a disaster. Randolph AFB is not as heavily involved as Lackland AFB, and some interviewees expressed that the facility has greater turnover in emergency management personnel. Civilians use Randolph AFB for additional manpower support, and staff from the AFB are involved in SNS planning and participate in the quarterly meetings mentioned previously. Some of the installation differences in community interaction may stem from geography; as noted previously, Fort Sam Houston is basically in the middle of San Antonio, while Randolph AFB is technically not within the city limits.

The VA is involved in civilian emergency management as part of the San Antonio Emergency Hospital Disaster Group and serves on the subcommittee for the Regional Medical Operations Center, as is typical for San Antonio–area hospitals. In addition, the VA is a major player in the local NDMS, but the public health personnel interviewed from local installations have not worked with the VA very much. The VA is a federal coordinating center, so it is linked into the NRF and DoD contingency plan this way.

Capabilities in the Military and Civilian Sectors

There are different views among military and civilian actors about how much the military has contributed when asked for support. One civilian interviewee noted that the military turned down requests for assistance in the community hurricane response because of equipment concerns, whereas military interviewees suggested that it is harder for the military to commit equipment-based assets than personnel. Other civilian participants noted that, even when the military was less involved in EOC meetings, it helped equip shelters and provided personnel to support evacuees from Hurricane Katrina.

Military capabilities were viewed more as augmentations to civilian resources than as unique resources for the community as a whole. For example, Air Force aircraft firefighting is superior overall, but, for structural firefighting and technical rescue, the city responders are more appropriate. The military had several local installation-level resources that were viewed as especially helpful to the local community, including disease surveillance, decontamination teams, helicopters, Lackland's runway, explosive ordnance disposal, Special Medical Augmentation Response Teams, aerospace medicine expertise, and trauma care. Randolph's in-place patient-decontamination team was mentioned as another potential asset, as was the National Guard's CST.

Military capabilities are limited not only by the need to provide support on base (in garrison) but also by the need to deploy assets to combat areas. The installations intend to be self-sufficient, but they may need civilian support for events, such as air shows, that involve a large influx of civilians. Several civilian interviewees identified additional capabilities offered by civilian agencies for military installations, including the SNS plan, communications infrastructure, an EOC facility, large facilities for storage and shelter, and a strong hospital coordination network.

Disaster Preparedness and Response Facilitators

Interviewees suggested that negative public perceptions of the response to Hurricane Katrina pushed both military and civilian stakeholders to ensure that Texas was fully prepared for future emergencies. Other interviewees also viewed Hurricanes Katrina and Rita as a reason for greater attention to emergency management, though not necessarily due to the negative publicity. After initially relying on a national team to help them with Katrina and Rita incident management, civilian emergency planning staff in the San Antonio area now have a local team that meets on a monthly basis and includes military personnel. The representatives establish incident management criteria as a group and break off into smaller groups for planning. WebEOC serves as a "common denominator" for event management and helps provide situational awareness across functions within a single military installation and for civilian agencies. One interviewee stated that the response to Hurricane Dolly revealed how far the community has come since Hurricanes Katrina and Rita.

In terms of formal mechanisms that facilitate preparedness, grant money enabled the purchase of equipment and is viewed as the impetus behind the city's weekly emergency management meetings. There are functional MOAs between civilian and military fire and police departments, and local hospitals also have MAAs with the military trauma centers, which the community can access just like any other hospital. Also identified as a facilitator was having a centralized office or organization to help plan and schedule exercises to ensure that grant and guidance requirements are met, plans are vetted, and key players can participate at some level.

Communications have played a major role in improving disaster readiness in the San Antonio area. As a civilian public health official explained, "Communication is really our forte here in San Antonio." The community avoids many interoperability issues than could limit communications by deploying communications trailers that obtain information from one system and relay it to a different system. The trailers perform a universal translation across radio frequencies so that it seems like all agencies are on the same channel. Prominent civilian agencies (e.g., public works, schools, police) share a common radio system. This reflects a significant investment in technology, which one interviewee placed at $45 million. Right now, the system works with military units as well, but this could change if the military does not keep the civilian community advised of changes to its communications systems, such as radio upgrades. However, not all interviewees agreed about communications interoperability in the San Antonio area: One civilian interviewee claimed that there was no interoperability between the military and civilian agencies.

As for other enabling factors in emergency preparedness efforts, there is a perception among those we interviewed that the NIMS and the ICS provide a basis for coordination and common terminology between military and civilian organizations, at least for those who have adopted the same standards. Participants also stressed that face-to-face meetings are essential to help key stakeholders get to know each other before the event and improve communications. Finally, a few key civilian personnel have security clearances, which makes sharing information with and receiving information from the FBI, Secret Service, and CIA feasible.

Disaster Preparedness and Response Obstacles

Participants also identified several obstacles in emergency planning and response. Although the military installations and San Antonio officials are all using WebEOC, they are using different versions of the software that are not connected to one another. WebEOC gives commanders a tool to look at organizations and ascertain information, such as bed counts and truck availability. Interviewees claimed that the old emergency management system was very manpower intensive but could do everything and had real-time data. They felt that it is hard to dedicate manpower to keep WebEOC current, especially with different security levels of information. The data are there, but the cost would be steep if some agencies had to upgrade their WebEOC software so that all sites could implement the same system. Interview participants expressed interest in more real-time capabilities and more GIS capabilities.

Interviewees mentioned other challenges that inhibit collaboration, such as legal and policy issues. Policy for DSCA, the Stafford Act, and Posse Comitatus were perceived as limiting the potential for military-civilian interaction in preparedness efforts. Lackland and Randolph interviewees specifically mentioned that they are authorized, through the Stafford Act, to provide certain forms of immediate assistance outside the base in the initial 72-hour period. But there is still a lack of clarity regarding DSCA-related procedures. For instance, military participants asked, "When is approval from a higher headquarters command necessary?" Additionally, during civilian interviews, the view was expressed that increased security on installations since the terrorist attacks of September 11, 2001, makes it harder for civilian stakeholders to enter the base and contributes to a feeling that the installations are isolated from the community. Furthermore, the military has a lot of resources, but restrictive policies that preclude personnel from sharing with community partners. Finally, the ongoing overseas contingency operations (i.e., the efforts in Afghanistan and Iraq) have caused supplies to be pulled from local military installments to ship overseas.

Competing demands also posed a challenge to preparedness efforts. Emergency management is often a collateral duty for military personnel, so it may not get the time and attention even they agree it deserves. For example, medical-care needs will trump PHEO position responsibilities. Moreover, other operational demands compete with resources for any planning efforts.

OPTEMPO is a related issue; for example, Lackland AFB personnel claimed that, at any given time, 25 percent of hospital staff is deployed. The civilian side also faces staffing constraints. For example, DHS is strongly encouraging San Antonio to establish a Fusion Center, but the San Antonio Police Department is working to ensure that more basic law enforcement responsibilities are handled appropriately. Coordinating the participation of multiple functions for the planning process can also be challenging, and it may be especially difficult, given both the aforementioned tendency for emergency management to be a collateral duty and the perceived lack of a mandate or directive from higher headquarters to make preparedness a priority.

Another obstacle to effective emergency planning is data sensitivity. Information sharing with public health officials is difficult because military officials perceive that civilian authorities are reluctant to release epidemiology-related data. Conversely, military personnel cannot reveal classified information to individuals lacking the required clearances and a need to know. Even when information is shared, military and civilian planners use different technical terms. The NIMS helps somewhat, but the different sectors in response have their own terminology, which causes confusion.

Finally, staff turnover threatened to weaken collaborative efforts. Although there is some continuity of emergency management personnel, especially at Lackland AFB, problems with coordination still exist because base leadership changes every two years. Also, additional changing and consolidating of leadership on the military side due to the Defense Base Closure and Realignment (BRAC) Commission may be hard for the community side to follow.

Decision Support Tool: Potential Application

Interviewees perceived less of a need for a decision support tool to reinvent threat or risk assessment processes and suggested instead that RAND build on or otherwise use tools already available. For example, Force Pro was considered a most useful tool to quantify risk and help an installation commander prioritize mitigation efforts. Respondents emphasized applications to enhance asset visibility.

At the time of our research, there was no current planning tool that included all assets. Air Force civil engineering and transportation are acquiring a system for personnel and equipment accountability. Air Force civil engineering has a 702 contingency response plan, which lists all available resources and how they are to be used in the event of contingency operations. However, there is no comprehensive database of assets.

Interview participants described the type of planning and event management support that would be useful to them, including the following features:

- Planning
  - a computer-based tool to do a hazard vulnerability assessment or prepare a response plan and SOPs and action plans that were tied to the hazard vulnerability assessment. All the organization or user would have to do is add telephone numbers, position titles, and site-specific information.
  - a template on how to identify resources and how to utilize them effectively
  - regulations, frameworks, and requirements all in one place with a comprehensive list and links. Ideally, conflicts in guidance would already be resolved.
  - operational checklists based on plans and SOPs, such as a shelter checklist tool developed by the CDC.
  - asset visibility and common operating picture (i.e., situational awareness). This system would include
    - resources, especially military resources, with names attached, so that stakeholders know not only what is available but whom to contact to request the resource
    - list of local individuals (as distinct from organizations) who may have specialized, highly relevant expertise depending on the event
    - military assets and equivalents. Civilian actors want the military to educate locals about potential assets and resource management and typing teams. Civilians would like to know what military resources are available that could be combined with civilian resources—and want to know who is responsible for that resource. It would be helpful to have the assets (equivalencies and capabilities) specified within a tool.
    - a standardized link that personnel can click to put information directly into system (e.g., bed reports)
  - pattern recognition, or automated data mining that would point out unique spikes as well as patterns in terrorist-related activity, spread of infectious disease, and other threats. The detect-to-treat needs can be moved to an earlier detection stage, which data mining could enable. If data mining cannot be fully automated, the system would provide a way to sort and prioritize information to aid in pattern detection.
  - grant coordination. Some grants allow the purchase of certain types of equipment. Coordinating grants across civilian players would ensure that the various ground response units do not overlap and that they meet all capabilities requirements.
- Event management
  - asset visibility and common operating picture. This system would be especially helpful for EOCs to view the current status of assets, provide one-stop shopping for resources, and integrate databases showing what rooms, equipment, and the like are available. This tool would ideally also have a current map of the base.
  - dynamic support managing demand for the SNS and other immunizations
  - greater GIS capabilities for mapping the evolution of a disaster (e.g., plume analysis) and for mapping response assets
  - ability to keep a log of events

  - means of sharing real-time information immediately, especially in an emergent situation (predominantly unclassified rather than classified data).
- Post-event
  - feedback results, be they from exercises or events, into a format that facilitates planning

Decision Support Tool: Implementation-Related Issues

Some organizations or actors may be reluctant to use a new decision support tool. Some Lackland AFB personnel felt that they have done emergency management work for a very long time and already use multiple tools (e.g., Microsoft NetMeeting, GIS), so they already know what to do and how to get things running. Accordingly, using a new tool or providing inputs to develop a tool was viewed to be of limited or dubious value. Furthermore, some emergency management representatives felt that emergency planning and response is driven more by people than by an electronic tool; thus, they would rather have another person than a new tool. For example, the state of Texas has a planning tool in its system, but it still needs the human element to make it work.

To implement a new decision support tool, one would need buy-in from all functions and organizations that would use it. Interviewees suggested that it would help to have an endorsement or policy directive from the federal level. Military personnel did not feel that there was a compelling authority for them to populate a tool that provides asset visibility beyond their installation.

Participants voiced other considerations for the implementation of a new decision support tool. The tool must be not only easy to use (perhaps include a built-in training application) but also easy to maintain in terms of data entry and other upkeep. Several interviewees commented on the lack of manpower for the sole purpose of inputting or managing data, which could be a time-consuming endeavor.

In a related vein, a civilian interviewee observed that a tool might not work because of the rapid changes in local capabilities and in leadership, including some stemming from BRAC. A new tool would need to be very dynamic and able to reflect those changes and keep the information current. It may help to have designated updaters of different data elements. For instance, an administrator can lock some assets on the system so that only a designated authority can revise information, and only with authority to do so.

Data privacy and classification were two other issues that interview participants identified as critical in implementing a decision support tool. Officials need to be mindful of protecting data about individuals as specified in the Health Insurance Portability and Accountability Act of 1996 (Pub. L. 104-191), the Freedom of Information Act (Pub. L. 89-554), and the Privacy Act (Pub. L. 93-579). This is especially relevant on the medical side and with respect to handling evacuees. If medical patients are involved, the system must have the ability to send and receive information encrypted, whether it is from civilian hospitals, the VA, or the military. Additionally, a decision support tool would need some sort of access restrictions, such as password protections or ways to mask and unmask data depending on a stakeholder's need to know. Public health department representatives expressed similar concerns regarding their data, including information such as population demographics, medical surveillance, biosafety level 3 (BSL-3) lab capabilities, and how the department plans to serve the community. The system must also be able to resist attacks from hackers and should have two servers for redundancy.

Finally, participants emphasized the need for interoperability. Existing databases do not communicate well. Interoperability is a problem within the Air Force, across DoD elements with other government agencies (such as the VA), and between civilian and military entities. The systems have different firewalls, and, at best, sequential meshing can occur. A new tool would need to have a common interface that helps bridge the interoperability gap, including firewalls.

Summary and Conclusions

San Antonio emergency managers are responsible for a vast expanse of land—tens of thousands of square miles that encompass 22 counties and extend all the way to the U.S.-Mexico border. There are also two Air Force bases (Randolph and Lackland) and one Army installation, including a medical center (Fort Sam Houston and BAMC) in and around the San Antonio area. While both civilian and military agencies take an all-hazards approach to planning, hurricanes are the most common disasters that occur and thus an important focus for planning and exercises. Generally, planning for emergencies is an annual process, and civilian and military exercises occur year round. Military personnel participate in civilian exercises, although they are not tasked in the plans, as it is not clear what capabilities they will be able to provide, despite active MOAs and MAAs between military fire and police departments and trauma centers, respectively. Based on interview accounts, the Army installation appears to be more actively engaged with its civilian counterparts than are the Air Force installations. The VA is involved in the area's Emergency Hospital Disaster Group and is a major contributor to the NDMS, as well as fulfilling its statutory obligations to DoD. As with many military installations, issues of data sensitivity, OPTEMPO, and personnel turnover were viewed as obstacles in local disaster preparedness planning and coordination. Overall, however, interviewees in San Antonio indicated that emergency planning and preparedness were robust, especially after changes made following the U.S. response to Hurricane Katrina.

Norfolk and Virginia Beach Metropolitan Area, Virginia (May 2008)

Background

Norfolk, Virginia, is 54 square miles with a population of approximately 234,400 people, making it the major city in the Hampton Roads area (southeastern Virginia). It is also home to the world's largest naval base (NS Norfolk) and the North American headquarters for the North Atlantic Treaty Organization. The region's only international airport is also in Norfolk. Virginia Beach, southeast of Norfolk, is 248 square miles with a population of approximately 425,260. It is predominantly a resort city, with hundreds of hotels, motels, and restaurants.

The area is home to three major naval installations: NS Norfolk, NAS Oceana, and NAB Little Creek. NS Norfolk is the largest naval complex in the world. It is home to the Commander in Chief, U.S. Atlantic Command, and the Commander in Chief, U.S. Atlantic Fleet, as well as other "type commanders." NS Norfolk has a population of about 56,000 active duty personnel and covers 3,980 acres. NAS Oceana supports the Navy's Atlantic and Pacific fleet force of strike-fighter aircraft (F/A-18 Hornet) and joint and interagency operations. NAS Oceana has a population of about 4,300 active duty personnel and covers 13,390 acres, with more than six miles of runways. NAB Little Creek is the major operating station of amphibious forces of the Atlantic Fleet. The base has approximately 7,900 active duty personnel assigned to it and covers 2,373 acres.

The Norfolk community includes the 16 jurisdictions that comprise Hampton Roads. The region has been recognized as a UASI region and has received grant support from DHS through the UASI program. The Norfolk area is also designated a metropolitan statistical area for the FEMA Regional Catastrophic Preparedness Grant Program. This grant has already provided $2 million in funding for preparedness planning. Additionally, Norfolk is an NDMS federal coordinating center. The mass casualty federal coordinating officer and the emergency planning department head at the NMC Portsmouth are responsible for receipt, triage, and referral of patients across the surrounding 20 hospitals. MMRS members indicated that much of their grant guidance focuses on picking projects in high risk areas and using the National Planning Scenarios for planning purposes, including multiple IEDs, aerosolized anthrax, major hurricane, and cyber attack. Overall, the Norfolk area receives fairly substantial funding for emergency preparedness, and officials have considerable experience in both disaster planning and response.

The Norfolk and Virginia Beach areas are coastal and thus vulnerable to hurricanes and flooding. In addition, manmade events (e.g., terrorist attacks) are a risk, given the concentrated presence of military assets.

Interview Profiles

The RAND research team interviewed personnel at NS Norfolk, the regional operations center (ROC), NMC Portsmouth, NAB Little Creek, NAS Oceana, and civilian authorities from Norfolk and Virginia Beach. At NS Norfolk, interviews took place with personnel representing emergency management, medicine, CBRNE, security, and the ATO. At the ROC, the RAND study team met with future-operations planners. At NMC Portsmouth, researchers conducted one large, lengthy interview session that included emergency planners as well as operations, security, and medical staff. NAB Little Creek interviewees included security, ATO, fire, medical, and emergency management personnel. At NAS Oceana, the RAND study team interviewed individuals responsible for fire, security, medical, and emergency management. In the civilian Norfolk and Virginia Beach communities, we interviewed an official responsible for mass casualty preparedness, MMRS, and infrastructure, as well as employees in public health, fire, police, medicine, and emergency management, and the city commissioner's office. Finally, the RAND research team also interviewed an emergency preparedness official at the Norfolk region VA office.

Definitions of Community

Civilians generally view installations as independent from the city. They communicate and work with the military, but disaster planning is not integrated. One notable exception to this is public health, with which military officials have been involved through SNS and MMRS activities.

Military definitions of the local community varied according to the type and scale of event, although the overlap was in the Hampton Roads area. The entire region plans for hurricanes, but, if a bomb were to go off at the local performance amphitheater, only Virginia Beach would be considered the community. If NMC Portsmouth is setting up a patient reception area for a disaster, then all areas within a 100-mile radius would be the community. Yet, another interviewee said the community was "everywhere" and stated that staff will go out to provide assistance up and down the coastline as needed.

Although the military views its main responsibility as within the fence line, the military view of what comprises its community varies dramatically according to the organization and chain of command of the different service components. For example, Norfolk is headquarters for Navy Region Mid-Atlantic, so its personnel have to consider a wider region. Also, the ROC is colocated at NS Norfolk, so the commanding officer of NS Norfolk might get direct requests from the local civilian community, because he also oversees the ROC. Technically, the area of responsibility for the ROC is from North Carolina to Maine, excluding the National Capital Region.

NMC Portsmouth, on the other hand, belongs to a completely different chain of command, through Navy Medical East, which is responsible for all Navy medical assets in the eastern United States. Navy Medical East also owns bases in Spain and Italy. The officials plan to set up a hospital command center, similar to a medical ROC. NMC Portsmouth and all of the clinics use a different computer system from that used by the rest of the local Navy installations. NMC Portsmouth interviewees also said that, since Norfolk is the NDMS federal coordinating center, the community includes the VA, Langley AFB, NAS Oceana, and NAB Little Creek. Furthermore, if NMC Portsmouth has to activate the federal coordinating center, the community would also include all the places where patients are dispersed.

The VA does not rely much on Fort Eustis or Langley to help, since, if the area was hit with a hurricane, Fort Eustis would be gone and Langley AFB personnel would already have evacuated. Accordingly, the VA plan is to rely on local civilian hospitals in the case of an emergency.

Community Risk Assessment

Threat Assessment. On the military side, NS Norfolk has developed its own program for threat and risk assessment, based on exercises and previous reports. The TWG decides the likelihood of the event and scores it A–E. The severity of consequences is scored on a scale of 1 to 5. The emergency management working group then evaluates these projections, using the formula likelihood × consequences = risk.

NMC Portsmouth uses a formal hazard vulnerability assessment developed by the Navy Bureau of Medicine and Surgery (BUMED). It is based on an approach quite similar to that reflected in the Kaiser Hazard Vulnerability Analysis assessment tool described in Chapter Six and presented in Appendix H. Assessment results change over time. Emergency planning officials make changes to account for tornado risks, though, generally, when they respond to something they use an all-hazards response approach and then scale down as needed. The assessment includes three threat categories: technological, manmade (intentional), and hazardous materials. In this tool, staff consider the probability of death or injury, then the tool provides a relevant threat percentage on the other end. NMC Portsmouth security also works with the local civilian community regarding threat assessments. Security personnel have contacted most of the neighbors around their perimeter to be on lookout for suspicious individuals. Some people have the security officer's cell phone number to call directly to report suspicious activities. Naval Criminal Investigative Service is responsible for doing frequent reviews of the current threat in Hampton Roads from an antiterrorism perspective only. NMC Portsmouth has subject matter experts who assess what issues are of concern and come up with a threat percentage.

In terms of surveillance, NMC Portsmouth personnel reported that they had a surveillance system with reporting from a number of civilian and military agencies. In addition, the local law enforcement agencies, including military security elements, receive daily email alerts from the Virginia Fusion Center. Suspicious activity Navy-wide is tracked and sent out immediately. NMC Portsmouth reports suspicious activity outside its gates via message traffic, but it does not report directly to local civilian authorities or any other entities. As soon as there is a change to the threat level, a message goes out across the community to all relevant actors.

NMC Portsmouth respondents noted that disease surveillance across hospitals and pattern identification is automated through Electronic Surveillance System for the Early Notification of Community-Based Epidemics (ESSENCE), although information has a 24-hour update delay. Every military treatment facility is required to have two people responsible for checking ESSENCE once a day. ESSENCE provides pharmacy data, hospital data, and other pertinent information. The chief medical officer will still hear about outbreaks long before ESSENCE issues an alert, because people get on the phone when something big is happening. In the worst-case scenario, NMC Portsmouth could call medical staff at Camp Lejeune, North Carolina, as they will also be aware of emerging outbreaks.

On the civilian side, each jurisdiction has to conduct its own risk assessment. The state also has a vulnerability assessment. Hampton Roads has a regional vulnerability assessment for all of the health-care facilities. A civilian emergency management representative discussed risk-based planning as a process of using a regional mitigation plan and then developing further plans and procedures from that. The focus of threat assessments has been hurricanes, HAZMAT incidents, and pandemic influenza. However, this interviewee described civilian situational awareness as "barren" and noted feeling out of the loop.

Vulnerability Assessment. Military vulnerability assessments are carried out by visiting Joint Staff personnel, who use the DTRA's JSIVA tool. The emergency manager at NAB Little Creek also carries out assessments for each installation, partly in order to help installations prepare for the JSIVA assessment. The JSIVA assessments feed into a field exercise that, in turn, leads to refinements in the preparedness process and further vulnerability assessment. The Joint Staff assessments tend to be more punitive than the local assessments.

The military installation ATO at NAB Little Creek does vulnerability assessments more frequently—monthly. The units will select a different vulnerability area, such as water or electricity, and assess the installation's preparedness in that area. Additionally, the military clinics at NAS Oceana and NS Norfolk conduct annual hazard vulnerability assessments and provide them to NMC Portsmouth superiors. The hazard vulnerability assessment is a spreadsheet that includes natural and manmade events and helps to score risk so officials can see the worst case. Then emergency preparedness staff can make the response to an instruction specific by adding a contingency plan for the worst risks. Hampton Roads also does a vulnerability assessment, but it is not aligned with that of NMC Portsmouth. A committee decides which vulnerabilities are most critical through monthly meetings featuring representatives from medicine, facilities, and finance. Either a regional representative or the chief naval officer (top-ranking uniformed Navy officer) visits the area yearly to conduct a vulnerability assessment. These particular installation assessments are combined into a consolidated vulnerability assessment, rather than focusing on separate areas of vulnerability, such as fire, terrorism, or medical.

In the VA-DoD contingency plan, the VA Emergency Preparedness Coordinator looks at the process for receiving military casualties from overseas for medical stabilization. To show what vulnerabilities are involved in that situation, the VA center timed the flights for how long it would take to get to Langley AFB or NAS Oceana, and then time from there to the final destination. Its staff also went to Richmond to look at the air strip, what it took to get the patients down to their area, and what could get in the way. The VA and DoD also have an emergency management committee to help identify and deconstruct vulnerabilities.

In the civilian sector, there is essentially little distinction between threat and vulnerability assessment. MMRS representatives report that each jurisdiction in Virginia has to go through risk assessment and that the state also conducts a vulnerability assessment at the emergency management level. As the regional MMRS, they rely on each jurisdiction's risk assessment.

Local Planning and Exercises

The local Navy installations use an all-hazards approach and hazard-specific annexes in its disaster preparedness planning and uses templates from the CNIC. Individual facilities will create a base plan with functional annexes and event-specific annexes. While the military approaches planning via all hazards, it also conducts specific exercises. Different teams have different exercises depending on their function. Once per quarter, all teams conduct an integrated exercise. Additionally, NS Norfolk has established MOUs for ambulance support.

In terms of civilian exercises, organizations seem to be well integrated with one another throughout the region. For instance, there is the Hampton Roads Emergency Management Committee and the Hampton Roads Emergency Management Technical Advisory Committee. The military is represented on both of the committees, but the extent of military participation in civilian exercises is limited. Despite civilian invitations to the military, many civilian interviewees commented that there has never been a good, integrated hurricane exercise. However, others mentioned participating in the annual state hurricane exercise. NMC Portsmouth meets with the local emergency planning committee, which features 15 people representing different agencies. Hampton Roads staff meet every month with delegates from 26 hospitals.

As for the specific types of exercise, MMRS planning focuses on mass care, shelter preparation, evacuation, and commodity and resource management for catastrophic events. Through the regional management technical advisory team, the MMRS mass care representative can reach out to all emergency managers on the civilian side. Some federal coordinating exercises on patient reception have also been held in the region as part of the NDMS.

One military emergency planner cited several previous tabletop exercises as well as a field exercise planned for June 2008, one month after the RAND study team's visit. The perception exists that getting all the players to the table at once, even installation players, can be problematic. There is not much engagement by fire and medical officials for planning tabletop exercises, and their participation is minimal. The military did not involve city officials in their own upcoming tabletop and field exercises because the city government is better prepared and the military wants to work out its own problems before expanding participation to include civilian counterparts. In fact, one official stated that he was told not to talk with the city because the military personnel were inexperienced and the Navy did not want them to volunteer something they were not authorized to offer. There was a sense that "the reins have been loosened," so, at the time of our research, military personnel could talk with city personnel but could not obligate anything. Other military officials stated that they had not exercised in several years, and one claimed that there is a lack of funding to participate in exercises; however, military staff do attend classes related to emergency management.

The VA has an all-hazards manual that is 600 pages long and includes drafts of plans, standards, a hazard vulnerability assessment process, and other features that cover all hazards, from fires to pandemic influenza to WMD. VA leadership was integral in developing guidance, which is revised and disseminated annually.

Key Actors and Coordination in Local Planning

Local military planners know that their immediate responsibility is to entities within military facilities—civilian, military, contractor. They provide support only within a 12- to 24-hour response time frame from the start of the event, so they are unable to engage with the civilian community in support measures for a long time. However, they acknowledge that this is a somewhat fluid restriction. There is interest in including the civilian community outside the military installation, but it is sometimes difficult due to these restrictions. The one exception is fire personnel, who have the most active agreement with the local civilian community. NAB Little Creek–based personnel expressed the desire to include the civilian community in their planning, and the commanding officer has invited the mayor, Virginia Beach police, and schools to the base, since the installation is in their city.

In contrast, several interviewees noted that some installation personnel do not consider it necessary to interface with local civilian personnel, thinking that interactions should be left to regional personnel and other Navy officials, such as Naval Criminal Investigative Service agents.

Civilian emergency planners have found it difficult to engage military leaders except in the functional area of fire and EMS. Local fire departments have automatic MAAs with military fire departments, which are negotiated at district or regional level. The civilian EMS offers ambulances to bases. Military firefighting personnel respond to incidents off-post. Since all firefighters train to the same standards and obtain similar credentials, it is easier for them to work together, although they are still using different radios. For EMS, the military provides less capability to augment civilian resources, but there is still communications and collaboration between the sectors, possibly again due to similar training and credentialing.

Otherwise, civilian interviewees claim that it is hard to engage military functions, such as security, emergency management, and public health. Civilian planners indicate that they have tried to reach out to the military but that, at some level, they know that military leaders have a different mission and cannot guarantee resources or assets. Civilians indicate that asset visibility would facilitate a more coordinated local emergency response. In addition, civilian emergency management personnel suggested that some joint training with the military would help to ensure that they are speaking the same language. Medical and public health interviewees also indicated that they have tried to collaborate with NMC Portsmouth with little success, except in the area of pandemic influenza planning.

Overall, military-civilian interaction is at a nascent stage. Military EOCs were set up very recently, so on-site installation coordination for emergency management planning was fairly new at the time of our site visit. Integration and collaboration is increasingly recognized as important, but more so for exercises at this juncture. The local actors, especially on the military side, have only recently begun moving toward more inclusive planning and exercises. During our interviews, we learned that it has been a challenge to push past functional stovepiping within an installation and to bring different functional actors to the table for more robust coordination. Military-civilian interaction seems to be greatest with the largest installation, NS Norfolk, although interviewees noted room for improvement. NAS Oceana and NAB Little Creek are smaller and have less direct relationships with the community. NMC Portsmouth seems to have a positive rapport with the public health and medical civilian community. The

VA representative claimed that the MMRS has really helped to bring the community together: "All of a sudden there was this money that we could [use to] coordinate this response and get hospitals working together."

Anything beyond local response goes up two separate chains of command: one for the military and one for civilians. These chains of command are more likely to meet at the top for coordination than to meet at the bottom, except in the case of immediate threat to life or property. Local augmentation of emergency response capabilities is sometimes coordinated via existing MOUs or MOAs, but larger-scale responses would likely follow reporting and requesting procedures. For the military, this means that the local installation emergency manager notifies the military's ROCs of its requirements. If military leaders need civilian assets, they may request support via the Secretary of the Navy or the state governor. On the civilian side, although local emergency managers and the ROC do sometimes get individual requests from local authorities for support, the military often tells them to use their own chain of command, (i.e., through the governor) to request military assets, especially for major events.

Capabilities in the Military and Civilian Sectors

On the military side, there is a perception that city authorities have more resources at their disposal than do the installations. Receiving grant money has helped civilian authorities to do such things as update communications systems (new radios, mobile communications vehicle) and invest in WebEOC. Many military interviewees felt that the civilian community has more capabilities to offer to the installations than vice versa. For example, the installation would need SWAT-team support from local police, NAB Little Creek lacks CBRNE equipment, and local civilian hospitals have a greater capacity for patient care, especially long-term care. One exception is the Navy's firefighting support. Naval fire department assets provide not only mutual aid to civilian communities but also automatic aid, which, we were advised, resulted in a number of military responses to fires outside installation boundaries.

NMC Portsmouth interview participants noted that civilian organizations do not request their capabilities because they understand the military's limitations. They indicated that they can see the civilian community's assets, such as beds, but did not believe the reverse to be true. NMC Portsmouth interviewees stated that they have no unique capabilities, aside from their CBRNE training being slightly advanced. They can quickly mobilize medical evacuation capabilities and call in patient transport, but civilians have Nightingale Regional Air Ambulance Service and other medical evacuation capabilities. If a hurricane hits Norfolk, NMC Portsmouth would use its airplane-landing facilities to provide medical evacuation out of the area, and DoD would take the lead. NMC Portsmouth has an MOU with the Portsmouth police SWAT team. NMC Portsmouth would need support for any type of explosive threat as well as Navy bomb and drug dogs.

NAB Little Creek personnel indicated that the installation could provide water trucks, tankers, and trucks that can transport debris and rubble out of a disaster area, but only if it is declared a federal disaster area. From the civilian community, NAB Little Creek noted that it would request CBRNE/HAZMAT support from the city, given its aforementioned limited capability. NS Norfolk can offer fire capabilities, EMS, explosive ordnance disposal unit detachments and a large HAZMAT response team. The installation also has MOUs in place with local mortuary units. However, none of the military capabilities that civilians discussed was regarded as a unique capability to the area as a whole. Furthermore, the civilian community's ability to provide SWAT capabilities is the only critical capability for which the base has need for civilian support.

Civilian planners recognized that they had more resources for fire and EMS, but it was not clear that they thought that resources were as lopsided as military interviewees claimed. The military would most likely request ambulances and staffing in an emergency, and civilian agencies, in turn, offer some HAZMAT capabilities and counterterrorism resources. In the civilian community, Sentara Norfolk General Hospital is the only major trauma center in the area, so all trauma patients would need to be transported there. Fire representatives believed that they could contribute technical teams; HAZMAT teams that are local, regional, and for the state; and EMS support. The civilian fire department also manages the CERT program, EMS provides ambulances, and both fire department and EMS assign staff to the local MMRS team. Local police representatives offered that they have two helicopters for the area, a bomb squad for the terrorism side, and a marine unit to assist the Coast Guard.

The VA is more constrained in terms of capabilities in a disaster. It is not the first responder if something happens at nearby Fort Eustis, and staff do not leave the medical center to respond to incidents at Langley AFB or Fort Eustis. The VA agreement with the city is that the city is the first responder and the city will contact the VA if it needs to. Most VA facilities do not have a fire or EMS department, so they rely on the city for help.

Disaster Preparedness and Response Facilitators

Despite legal restrictions on deploying installation personnel to assist civilians beyond the "immediate" time frame, the military can get around many of these issues through base "volunteers": Essentially, staff are sent home from the base and volunteer as civilians. The ROC was noted as a resource for coordinating asset requests from both military installations and civilian agencies, and both WebEOC and MOUs were perceived as enablers to sharing information, personnel, and supplies.

MAAs are another crucial facilitator of planning and response. Current MAAs are primarily between military and civilian fire and EMS departments. In terms of planning and training augmentation, participants cited the NRF and FEMA training course on incident command. The Virginia Hospital and Healthcare Association has a program used by several but not all local hospitals. It is updated daily by civilian medical staff, and it posts drill schedules and other status information.

Having multiple avenues for communications can also promote preparedness and response activities. Some routes for information sharing cited during interviews include the previously discussed WebEOC, as well as very high frequency (VHF) radios that provide a common military-civilian interface. Hampton Roads officials purchased VHF radios so all hospitals could communicate, and interview participants thought that there was decent military-civilian crosstalk. Interoperability with the VA was a problem until about 2006, but, at the time of our research, since the VA is part of the area's hospital community, it has its own radio tower, through which employees can access information on bed counts and resource shifting. Informal communications between sectors is also necessary. One interviewee claimed that 90 percent of emergency preparation is networking and getting to know the people one will need to call upon in an emergency.

Disaster Preparedness and Response Barriers

One challenge to effective emergency planning is the legal restriction on military engagement—i.e., military personnel cannot respond outside their normal duties unless, as one interviewee put it, the purpose is to "save lives or prevent immediate property damage." There are also limitations to carrying military weapons off the base and getting the military to release a list of emergency managers to the civilian sector.

Financing can also be problematic in disaster preparedness. Military interviewees commented that localities do not like to ask for military support because services cost more if they have to be reimbursed than would similar assets borrowed from another civilian community. Normally, the city or state government covers the costs of assets that have been borrowed. Even military-to-military sharing is problematic, and interviewees shared anecdotes about having to make 50 phone calls and pay $6,000 per hour in fuel for borrowing a C-130 for a drill.

Communications interoperability continues to be a major obstacle for both the military and civilian sectors. For instance, the fire department has City of Norfolk fire radios, but police and security do not, and personnel must go through a dispatcher to communicate. The Enterprise Land Mobile Radio system has been touted as a solution to this problem, but it still has not been implemented. Sometimes, officials within the same branch of the armed services cannot even link easily. In the Navy, for example, personnel can email one another directly but cannot send notification pop-ups to desktops. Furthermore, confusion arises because of different terminology among disciplines and different coding systems.

Staff continuity is another challenge that impedes progress in emergency planning. Without staff connections and points of contact intact, interviewees noted that it can be difficult to accomplish long-term projects and strategic planning. One site has not exercised in six years, due at least in part to employee turnover. Additional challenges include time and money constraints to do large-scale drills and problems with physical access to installations, since some of them would be closed off in the event of a disaster. For example, NMC Portsmouth has only two main access roads. Access to the medical center is via bridges and tunnels that would be vulnerable in a major event and prevent employees from reporting to duty, as well as responders from getting to victims.

Decision Support Tool: Potential Applications

Interview participants suggested the following characteristics and capabilities for developing a decision support tool:

- General
  - Planning has several phrases, and each phase could use different tools:
    - six months prior (the "ideal")
    - 0–24 hours after event
    - 24–72 hours after event (period during which one can assess damage, assess how much existing assets are being used, and determine where the next closest assets are located)
    - 72+ hours (federal support available; overlaps with prior period)
- Planning
  - knowledge repository. Synthesize relevant passages of various DoD, service, and non-DoD publications, as well as pertinent lessons learned (throughout the continental United States) so that all useful information to inform planning or a response (espe-

cially for specific scenarios) would be at the individual's fingertips, presented in a manner that avoids cognitive overload.
    - guidance documents. There are many instructions, so it would be nice to have software allowing one to click on a topic and see all relevant references from each of the 80 references instead of reading them all one by one.
  - planning templates, checklists, and standard operating procedures. These documents would walk people through a logical planning process for EM that would include checklists, such as some of the existing programs or lists of types of individuals for whom contact info is critical, or generate lists of SOPs. Templates would also facilitate quick communications, as officials would not have to constantly rewrite messages, and help reduce paperwork.
  - shortfall identification. This process would combine threat vulnerability assessments with budget planning. One could input everything, then have a financial component to help calculate costs for the assets needed—the output would be level of risk and resources needed to address that risk.
  - surveillance and pattern detection. Information sharing across different locations and organizations would facilitate coordination and effective response.
  - risk assessment. A technological process should be in place for risk assessment across the region.
  - communications and points of contact. A tool should enhance communications and put together resources more quickly than looking through books and telephone lists.
  - interoperability. A decision support tool should have the ability to tie into existing systems, such as C4I Suite, Computer-Aided Management of Emergency Operations (CAMEO), Areal Locations of Hazardous Atmospheres (ALOHA), NWS, and WebEOC.
  - centralized database or tool. All data sources from all sectors can feed into a centralized database or tool to create a system in which officials can access information, such as the structural integrity of a building, that may not normally be available in current emergency response systems.
- Event management
  - asset visibility. Civilians in particular wanted a tool to help them put in a mission request, be able to contact bases, know which bases can provide a resource, and know the decision time to get that asset. A request form scripted by the military for civilians would help ensure that requests do not get rejected.
- Post-event
  - Have AARs available.
  - Documentation of AARs and other reports might help with the problem of turnover, which is a challenge to the sustainability of protocols, especially for disaster preparation, since it is a collateral duty for many.
  - The tool could include past plans or reference pages to see what has been done—not just scenarios but events that actually happened, such as the medical response to Katrina or the response to the earthquake in China.

Decision Support Tool: Implementation-Related Issues

Interview participants discussed some challenges in implementing a new decision support tool. First, the tool would need to be Web based to avoid hardware installation; easy to use, since it may not be used frequently; and affordable, so agencies will agree to the investment. Classification and privacy issues are also important to consider, and different organizations may have different security restrictions on installing programs or accessing websites. The tool should also address concerns of interoperability among disparate communications systems.

Summary and Conclusions

NS Norfolk is the world's largest naval station, both in terms of acreage and personnel. The cities of Norfolk and Virginia Beach work through the city manager and, to a lesser extent, through the mayor to decide on planning resources and to activate protocols. There is considerable military and civilian emergency preparedness under way, and interview participants believe that there are still opportunities to enhance local coordination.

City of Columbus and Muscogee County, Georgia (July 2008)

Background

Columbus is a large city in an otherwise relatively rural area. Its population of 188,660 (2006) is third largest in Georgia (total population 9.4 million), and many of the surrounding rural counties depend on it for resources. It has a consolidated government, meaning that the city and the county in which it is located, Muscogee County, have merged to provide many services to its citizens (e.g., public health, public safety) more efficiently (Columbus Consolidated Government, undated). Although there is both a city police department and a county sheriff's office, there is only one office for homeland security, one agency for fire and EMS, and one agency responsible for a district of Georgia that encompasses the city, Muscogee County, and beyond.

Most of the civilian emergency management staff in Columbus have either lived in the area for a long time or hail from the area, have at least five to ten years of experience in Columbus itself, and have served mostly in comparable positions throughout their careers. Military leaders have a similar level of functional expertise but, with the exception of the Fort Benning fire chief, had been in the Columbus area for less time. This was particularly the case for active duty personnel tasked with emergency management responsibilities at the installation. Civilian emergency managers were quite familiar with the military, and many individuals with prior military experience have made their way into the local civilian workforce. Perhaps most notably, the previous mayor, a former city manager, and an official at the public health department had all served in the armed forces.

While it may seem surprising because of the small size and arguably low threat profile of Columbus, the city has received preparedness grants and funding for years—Columbus was a Nunn-Lugar-Domenici city even before 2001. The ability of city officials to gain resources despite being less in the public eye than larger cities was suggested to be due in part to the talents and political pull of some key stakeholders. In addition, there is arguably a need for Columbus to serve as the nucleus for support within southwestern Georgia and beyond to Alabama, Florida, or Mississippi. In addition, some stakeholders contend that having a large, high-profile Army installation within the community, along with prominent businesses (such as American Family Life Assurance Company of Columbus's [AFLAC's] world headquarters, means that there is a heightened risk of terrorist attack in the local area. Funding through the MMRS and Nunn-Lugar-Domenici programs has helped the community stay ahead of the curve in terms of training and equipment such that, even before the September 2001 terrorist attacks, Columbus was already prepared to respond to events like an anthrax scare.

Fort Benning, with a population of nearly 22,000 active duty personnel and covering 171,873 acres, is home to the U.S. Army Infantry Center and School and the Martin Army Community Hospital. Overall, Fort Benning is the sixth largest military installation in the United States and has the third largest troop density. Fort Benning trains almost 120,000 soldiers a year, with an average of 14,000 soldiers on base per day. In addition, as a result of BRAC, the U.S. Army Armor Center and School will relocate to Fort Benning by 2011, raising its population by 30,000 people. Fort Benning is also home to the U.S. Army Airborne School (used by all four services) and the U.S. Army Ranger School. Deployable and deployed units stationed at Fort Benning include the 3rd Brigade Combat Team, 3rd Infantry Division; 75th Ranger Regiment; Western Hemisphere Institute for Security Cooperation; 14th Combat Support Hospital; 13th Combat Sustainment Support Battalion; 11th Engineer Battalion; 209th Military Police Detachment; 789th Explosive Ordnance Detachment; 24th Ordnance Detachment; 63rd Engineer Company; 233rd Heavy Equipment Transportation Company; and the 988th Military Police Company (U.S. Army Maneuver Center of Excellence, 2009).

The primary disaster experience for the Columbus region involved providing support for other areas affected by hurricanes and tornadoes. For instance, the Muscogee County's Sheriff's Office helped with public safety in Mississippi during Hurricane Katrina, and local medical personnel worked to ensure continuity of medical care after a hospital in Americus, within a rural, neighboring county, was destroyed by a tornado. There have also been incidents that affected the area directly, such as tornadoes, Hurricane Opal, and an emergency plane landing that had the potential for mass casualties. Additionally, an annual Western Hemisphere Institute for Security Cooperation (formerly, the School of the Americas) protest at Fort Benning draws tens of thousands of protesters to the city every November, and the civilian community and military both prepare for this event, which first took place in 1990 (SOA Watch, undated). However, interviewees suggested that hurricane evacuation is the main threat for which the local community regularly plans and prepares.

Interview Profiles

The RAND research team conducted nine interviews in the Columbus area: four at Fort Benning and five with staff from Columbus/Muscogee County. Interviewees at Fort Benning included representatives from public health, antiterrorism/intelligence, public safety, fire, CBRNE, and emergency management. Civilian interviewees included those responsible for emergency management, fire, public health, public safety, and antiterrorism for Columbus/Muscogee County, as well as liaisons from the mayor's office and a representative from the VA. RAND staff were unable to interview budget, operations, or executive-branch employees.

Definitions of Community

Most civilian and military leaders agree that, geographically, the community includes at least the tri-city area of Columbus (Georgia), Phenix City (Alabama), and Cusseta (Georgia)—roughly the Columbus, Georgia–Alabama metropolitan statistical area. However, military leaders qualify that their first responsibility is inside the gates of the base even though they will offer assistance to civilians if they can. Accordingly, Fort Benning personnel tend to have a narrower view of community, focusing primarily inside the installation's boundaries and on the nonadjacent Army enclaves for which Fort Benning is primarily responsible. At the same time, there is recognition that many events that affect Fort Benning would also affect the community around it, and the Army is prepared to help the community when and how it can.

Civilians' definition of community was more expansive, not only extending to Columbus but often including many of the neighboring counties as well as municipalities in Alabama, such as Phenix City. One interviewee even referred to the entire metropolitan statistical area as the community. Depending on the nature of an event, Columbus-centric personnel may need to provide support to these areas. Civilian leaders are well aware of the constraints on military response and noted that they temper their expectations accordingly.

The VA perspective of the community boundaries differed somewhat. VA hospitals are located in Montgomery and Tuskegee (Alabama), with only outpatient clinics in Columbus. Thus, VA emergency management staff focus more on central Alabama—Montgomery and Macon counties, with a limited role in Columbus. The VA medical facility is located closer to Fort Rucker and Maxwell AFB, and it has minimal interaction with Fort Benning.

Civilian and military interviewees in two particular functional areas—public health and homeland security and intelligence—tended to have the broadest view of what constitutes community in their area. For example, military public health personnel referred to participation in the local MMRS and noted that the community includes not only active duty personnel but also DoD civilian personnel, most of whom did not live on the post. Public health employees also took wide view because theirs is a regional health department, formally responsible for a large number of cities and counties in southwestern Georgia, as well as part of the local MMRS. Homeland security and intelligence personnel also had an inclusive perspective on community, especially from the standpoint of intelligence gathering and threat assessment. One interviewee in this area felt that the local community comprises anyone with a vested interest in the jurisdiction, not only residents and public-sector agencies but also tourists and local businesses.

Community Risk Assessment

Threat Assessment. Threat assessment processes and tools vary greatly, not only across military-civilian boundaries but within them. For example, Fort Benning intelligence analysts use the DTRA guidance and consider data furnished by the JTTF; Martin Army Community Hospital uses a hazard vulnerability assessment matrix in which events are ranked based on a combination of probability, risk, and preparedness, and CDC and Transportation Security Administration data inform the process. Columbus emergency management officials use expert judgment in which threats are a function of possibility, probability, and consequences, while another Columbus interviewee touted the city's use of a "hybrid approach" that relies on a human element and includes a matrix based on different categories of uncertainty. Public health officials use formal guidance, the NIMS Compliance Assistance Support Tool (NIMSCAST). The NIMSCAST is designed for the emergency management community as a comprehensive self-assessment support tool. Lastly, the VA uses the Kaiser Hazard Vulnerability Analysis assessment tool, with a risk algorithm in which anything over 20 percent deserves attention in the formal planning process. In one common theme, the Consequences Assessment Tool Sets (CATS) (SAIC, undated) was mentioned by military interviewees and includes Computer-Aided Management of Emergency Operations and its component elements (Areal Locations of Hazardous Atmospheres and Mapping Applications for Response, Planning, and Local Operational Tasks, or MARPLOT), Hazard Prediction and Assessment

Capability, and Chemical Biological Response Aide (CoBRA®) IV first-responder software systems.

Some of these differences in processes and tools stem from either a focus on different threats or analysis conducted by different functions. Such is the case at Fort Benning, where different threats are generally assessed by different parties: One person or organization addresses CBRNE, another focuses on public safety, and health officials at Martin Army Community Hospital conduct their own assessment. Military personnel responsible for antiterrorism and military intelligence conduct threat assessments for manmade (intentional and unintentional) activities that could have negative consequences for local personnel or facilities, including CBRNE, but they do not typically evaluate weather or health-epidemic hazards. However, Fort Benning personnel responsible for intelligence analysis reported using a DTRA framework, which does include a medical piece that extends beyond purely biological terrorism, such as an anthrax attack. Revisions to local threat assessments are done annually.

On the civilian side, multiple actors may be involved in the same assessment. For instance, the critical infrastructure planning committee has 40 private organizations, such as businesses, nonprofits, and religious institutions, which collaborate with law enforcement, fire, emergency management, and military. The VA, on the other hand, uses the Kaiser Hazard Vulnerability Analysis tool for risk assessment. The frequency with which assessments are conducted seems to vary on the civilian side and was less clear to the RAND team for some civilian agencies than for military entities. Further, interviewees viewed differently the benefits of having experts do assessments pertaining to their area of expertise versus cross-training people so that they have a more global view of all possible disasters and events.

There is little interaction across military and civilian lines to assess threats jointly. There is some sharing of results post-assessment, but there does not seem to be much up-front collaboration. However, local civilian officials participate in Fort Benning's FP working group and TWG meetings, and the military contributes to the Critical Infrastructure Planning Committee's threat assessment.

Vulnerability Assessment. At Fort Benning, staff conducts an annual installation examination based on the threat assessment, which looks at critical functions and infrastructure to identify vulnerabilities. Each potential vulnerability area, such as food, water, and communications infrastructure, is assessed by a different analyst. While military medical personnel rely on intelligence analysts and the ATO for terrorism-related aspects of threat assessment, medical personnel seem to play a greater role in the vulnerability assessment. On the civilian side, organizations use the vulnerability assessment in tandem with the threat assessment. The threat analysis is probability based, whereas the vulnerability analysis is consequence based. In the vulnerability-assessment phase, civilian officials consider the significance of consequences and isolate points at which the emergency would exceed staff capabilities. The public health department in particular mentioned using the NIMSCAST system for this purpose. Federal actors, such as DTRA are involved as are academics, including researchers from Texas A&M University who visited the area to assess community vulnerabilities in such areas as the water supply.

As for the VA, it again uses the Kaiser Hazard Vulnerability Analysis tool for this task and works independently from Fort Benning personnel and Columbus-area civilian officials. The VA relies on its hazard vulnerability assessment tool for pinpointing specific weaknesses, but, generally, its two hospitals are the most vulnerable because they house most of the personnel.

Local Planning and Exercises

Most if not all interview participants indicated that their plan encompassed all hazards and all emergencies; such plans tended to consist of a short overarching vision supported by a large number of event-specific or function-specific annexes. Planning efforts consistently incorporated natural and manmade events and were expanded to include terrorism as part of an increased emphasis on homeland security. However, plans for different functions and events were often drafted by different individuals, not all plans follow a specific process or any sort of set guidance, and these plans were not necessarily combined into a consolidated local civilian "master plan."

Fort Benning focuses on the installation in its plans but is also mindful of ways to assist the local community in a disaster. Civilian leaders share their plans with the military, as there is a push to engage them in discussions. However, civilians do not delegate any emergency support functions to military personnel. Rather, an overarching all-hazards/all-emergencies MAA broadly outlines the nature of support between civilian agencies and Fort Benning, function-specific agreements further refine some of the conditions for first responders, and requests are made on a case-by-case basis; no automatic aid agreements are in place.

Checklists and call-down lists of personnel needed for emergency response are major features of VA emergency plans. The VA also conducts various exercises with other entities in Macon County (Alabama). Its role during a disaster is limited to triage support because the facility cannot provide trauma care. VA staff would work with others to transport patients to hospitals to Elmore County or Lee County (both in Alabama). Representatives from Maxwell AFB are also involved in emergency planning and exercises.

Like their military counterparts, civilian representatives also conduct both tabletop and full-scale exercises. Exercises are both function specific, such as the mobile field force and SWAT-team training organized by local law enforcement agencies, and multifunctional. The latter type tends to be initiated by the city's emergency management division (within fire/EMS). Fort Benning personnel participate frequently in civilian tabletop exercises. Civilian and military personnel also exercise together in preparation for the annual School of the Americas protest; although both Fort Benning and civilian agencies have responsibilities independent of the other for this event (e.g., civilian entities monitor food and water issues downtown), other aspects of the protest require collaboration. Plans are not only revised as a result of these various exercises but also reviewed and revised usually on an annual basis. However, many are largely unchanged from when first drafted, except for updates to contact information and capabilities.

Key Actors and Coordination in Local Planning

For Fort Benning, the primary military actor in the Columbus area, disaster preparation plans are coordinated by a central planning office responsible for the majority of all Fort Benning's plans (i.e., not only those pertaining to emergency management). The process begins with a threat assessment that informs the development of plans. The plans, in turn, lead to exercises intended to test the plans, and revisions to the plans are made in response to deficiencies the exercises reveal. Stakeholder input informs the process throughout. Fort Benning conducts at least one full-scale field exercise per year, and its personnel participate in tabletop exercises organized by civilian agencies at least on a quarterly basis. Soon after the RAND study team visited, for instance, military health personnel planned to contribute to an MMRS-initiated tabletop exercise involving a tornado-like event off post. It had four objectives, one of which was patient tracking. Within the installation, exercises are a multidisciplinary process that includes fire, EMS, and other key responders, as was the case in a March 2008 exercise that featured WMD, hostages, mass casualties, and a chemical explosion, and function-specific training and drills are conducted throughout the year as well. For example, Fort Benning's fire department organizes its own CBRNE exercises and ensures that its personnel meet the national firefighting training standards.

Similarly, civilian planning engages a variety of actors, such as staff from public health, hospitals, fire/EMS, and public safety, in a process consisting of plan generation, testing, refinement, revision, exercises, and additional revision. Their plans are lengthy and feature many annexes, just like those of other communities.

Capabilities in the Military and Civilian Sectors

The RAND team gleaned relatively little information about joint capabilities-based planning. Limited data suggest that, to the extent it occurs, it is an ad hoc process on a case-by-case basis. Exercises and drills are seen as a helpful way of making different agencies aware of one another's resources and functioning.

There are few capabilities unique to one sector; usually, the situation is one of augmenting capabilities the other sector already possesses. Nonetheless, there are some assets that are relatively unusual or well-honed by Fort Benning, including explosive ordnance disposal and rotor-wing airlift capabilities; unique capabilities on the civilian side include river SAR, septic-system inspections, trauma care, orthopedics, and neurosurgery. In some cases, Fort Benning relies on the civilian sector for capabilities that entail niche skills from personnel or require a long-term, cohesive team (such as a SWAT team) that are hard to maintain at Fort Benning, given the nature and frequency of soldier rotations. In addition, given the civilian fire department's cache of CBRNE and HAZMAT equipment and the new CBRNE equipment acquired by the sheriff's office, civilian authorities may have more capabilities in this area than does Fort Benning. Civilian emergency management and homeland security offices also offer reliable communications and coordination capabilities with all agencies, including law enforcement and fire departments. Public health staff cited issues with a shortage in nursing-related capabilities but generally can cover most epidemiological tasks. The VA maintains a volunteer-management database as part of its own disaster preparedness and response efforts.

Disaster Preparedness and Response Facilitators

Several formal mechanisms enhance disaster preparedness and response activities in the Columbus area. There is the aforementioned umbrella MAA between the military and civilian sectors for all hazards/all emergencies, which is supplemented by more specific agreements between functionally equivalent first responders. Additionally, the aforementioned grants (MMRS, Nunn-Lugar-Domenici) and other sources of funds have helped the civilian community in terms of both equipment and training. The funding has enabled the civilian organizations to obtain equipment that promotes communications interoperability, such as mobile communications vehicles, and to establish a new SAR team. The planning and exercise requirements on which these grants are contingent have also facilitated positive civilian-civilian and military-civilian relationships. Finally, interviewees believed that emergency management had progressed due to the NIMS command structure, which both sides generally follow. Civilian interviewees noted that Fort Benning personnel are increasingly working to align with the NRF and its emergency support functions.

There are also informal factors at work that promote emergency preparedness in the Columbus area. Social connections were not only a facilitator but possibly an essential ingredient to effective disaster response. Interviews suggest that the relationship between military and civilian actors is longstanding, friendly, and respectful. They acknowledge each other's limitations in responsibilities but try to ensure that each sector has some awareness of the other's plans, capabilities, and actions taken during an incident.

Disaster Preparedness and Response Barriers

Interview participants noted several obstacles that hinder disaster preparation and response and identified solutions they use to circumvent them. One issue that arose was legal limitations imposed on Fort Benning by DSCA, a DoD policy, and Posse Comitatus, a federal law that restricts what law enforcement activities military personnel can engage in off post. Language and semantics also proved challenging, as many codes and acronyms do not translate across military-civilian lines or across functions. One strategy to overcome language barriers was having people with a military background on staff at civilian agencies. The recent move to "plain talk" has led to some improvement in communications.

Another difficulty interviewees observed centered on technical aspects of communications interoperability. Various organizations have purchased communications equipment from different vendors, and some of the smaller communities in neighboring rural counties have older equipment that is sometimes substandard. To combat these challenges, city officials have acquired state-of-the-art equipment, including machines like Raytheon's ACU-1000 interconnectivity system (a universal translator across frequencies), and the Muscogee County Sheriff's Office has deployed a mobile communications vehicle. Funding in the amount of $350,000 will be required to sustain the RAND Terrorism Incident Database through fiscal year 2010. However, interviewees believed that agencies still need to have people working as liaisons to compensate for communications-system deficiencies. Furthermore, the local public health department and Fort Benning's military hospital used WebEOC for disaster event management, but the local civilian hospitals relied on LiveProcess, a competing type of software (see Chapter Six and Appendix H). This means that, during an emergency, telephone calls are still necessary for such tasks as bed counts and finding available blood units. To address communications gaps between the civilian and military sectors, participants noted the use of communications-focused drills and redundant communications systems (e.g., BlackBerry® devices, pagers).

Other complications mentioned by participants include the lack of DoD-issued security clearances on the civilian side, which can hamper military-civilian intelligence sharing. In response to this issue, some civilian emergency management personnel have obtained relevant security clearances, which have afforded them greater access to work on counterterrorism efforts. Additionally, resources and scheduling constraints can hinder large-scale and multifunctional exercises; while personnel often want—and intend—to engage in field exercises arranged by another organization, sometimes, such plans are overcome by real-world events or other demands. Finally, military personnel turnover due to permanent changes in station and deployments can render it difficult to sustain institutional disaster preparedness knowledge and capabilities on the base and keep civilian agencies apprised on a reliable basis of who is in charge of particular functions. Using DoD civilian personnel at Fort Benning provided some continuity in emergency management and was one way to mitigate the effects of the unavoidable turnover of active duty personnel.

Decision Support Tool: Potential Applications

Interview participants suggested the following characteristics and capabilities for decision support tools to aid disaster planning and response:

- Planning
  - regulations, frameworks, requirements all in one place and linked, with conflicts in guidance resolved
  - a checklist function for planning
- Event management
  - pattern or trend analysis (e.g., State Electronic Notifiable Disease Surveillance System [SendSS] or ESSENCE on the medical side). There is a general interest among actors in connecting the dots across potentially related events, or "real-time data mining."
  - credential access for first-responder personnel, especially medical
  - ability to check resources in and out for use during an incident
  - health-specific functions, such as a dosage calculating tool
  - common operational picture—something like WebEOC or the Fort Benning Incident Operation Center's version of SharePoint®—for situational awareness
  - mapping technology to locate threats and monitor critical infrastructure
- Post-event
  - Be able to download actions into AARs.
  - Facilitate sharing reports and best practices.
  - Enhance capabilities for analysis.

Decision Support Tool: Implementation-Related Issues

Interviewees identified some possible implementation challenges in adopting decision support tools. For example, many employees are accustomed to relying on hard-copy planning tools, such as field manuals, NIMS boards, and SOPs, and are accordingly less familiar with using computer-based technology. They also note that computers are not always accessible or connected during an emergency event. In addition, interview participants stressed that staff need to be trained to use technology and that operating systems across different agencies need to be integrated. They suggest weaving the decision support tools into daily functions rather than just emergencies, which would improve familiarity and competency with the new tool. In a related vein, interviewees emphasized that decision support tools need to be intuitive and easy for all actors to use (i.e., "plug and play"), though it may be challenging to create something that works across installations. Finally, interviewees enumerated some key functions that a decision support tool ought to contain, such as portability (for use on site), password protection, automation, and minimal upkeep, that would facilitate its implementation and ensure its ongoing use across functions and across sectors.

Summary and Conclusions

In the Columbus area, much of the emergency planning and response activities relies on strong interpersonal relationships, and there was a clear sense among interviewees that ties run deep in this region. The short MAA that guides interactions across agencies was described by some of them as a "gentlemen's agreement." Interviewees asserted that there is no substitute for developing and maintaining relationships and that one of the benefits of joint planning and exercises is the opportunity for various actors to get to know one another and nurture personal connections.

City of Tacoma and Pierce County, Washington (September 2008)

Background

Tacoma is Washington's third largest city, located administratively within Pierce County and geographically at the foot of Mount Rainier and along the shores of Commencement Bay. The city has a population (200,000) about one-fourth that of Pierce County (800,000), and it occupies about 50 square miles of the county's 1,806 square miles.

Both Fort Lewis and McChord AFB are located in Pierce County. Fort Lewis is home to I Corps, with primary maneuver units being the 1st Brigade/25th Infantry Division and the 3rd Brigade/2nd Infantry Division. Fort Lewis is also responsible for Yakima Training Center in eastern Washington and home to Madigan Army Medical Center. Fort Lewis itself covers 86,000 acres, with an additional 324,000 acres at Yakima Training Center. Fort Lewis is home to approximately 22,000 active duty soldiers (see Table 4.1 in Chapter Four); however, not all of these individuals reside on the base.

McChord AFB is the home of combat airlift, flying continuous combat airlift and aeromedical evacuations in support of Operations Iraqi and Enduring Freedom and other contingencies around the world. Roughly 4,000 active duty personnel are assigned to the base, which covers 4,639 acres. As a result of BRAC, McChord AFB and Fort Lewis will work together more closely, consistent with a joint basing concept. This includes Madigan Army Medical Center handling the bulk of McChord AFB's medical needs, including EMS, and has motivated Fort Lewis and McChord AFB to interact with one another more for emergency management than was previously the case.

Threats to Tacoma, Fort Lewis, and McChord AFB are predominantly natural, including flooding, forest fires, severe weather (tsunamis, wind, and ice storms), volcanic activity, and earthquakes. The area has also experienced manmade threats, including a train derailment in Fort Lewis and occasional homemade bombs crafted by hunters and anglers. Local emergency managers also pay attention to agricultural and health-related hazards, as there was an incident of bovine spongiform encephalopathy (mad cow disease) in eastern Washington. Other recent hazardous events have included an earthquake in 2001; a severe windstorm in 2008 triggering a major, multiday power outage; and the December 2007 crash in Pierce County of a Blackhawk helicopter from Fort Lewis. This last incident raised issues about how to secure the area and prompted many AARs.

Interview Profiles

The RAND team conducted 13 interviews with a total of 42 civilian and military personnel in the Pierce County/Tacoma area. The 11 civilian interviewees were with persons representing emergency management, security or law enforcement, the fire department or EMS, health/medical and CBRNE/HAZMAT, and the VA. Interviewees from Fort Lewis and McChord AFB included personnel representing these same areas on the military side.

Definitions of Community

Generally, interviewees considered Pierce County as their community for emergency planning purposes, although Fort Lewis and McChord AFB gave priority to their respective installations, and the city of Tacoma narrowed its focus to its specific constituents for emergency planning and management. At the time of the interviews, civilian responders and managers had MAAs with surrounding counties; however, Seattle was not typically engaged in standard emergency management planning efforts (even though Tacoma law enforcement has assisted Seattle in such instances as the World Trade Organization protests in 1999).

VA representatives did not define community in the traditional sense of boundaries, but rather indicated that, in the case of an emergency, the VA would work through the command hospital (one per county) to receive notification of patients whom the VA would try to help.

Interviewees expanded their definitions of community when describing other areas of operations. For instance, McChord AFB is responsible for air evacuations worldwide, and its firefighting brigades have been deployed as far away as Idaho and California. There were also tests of regional civilian communications systems, which found that interoperability extends as far south as Portland, Oregon, and as far north as the Washington-Canada border, proving useful in a large-scale emergency.

Fort Lewis and McChord AFB personnel recognize that events that affect Tacoma/Pierce County will likely affect the installations, especially in the case of natural disasters and public health threats. Overall, civilian interviewees observed that their military counterparts have broadened their view of community. This is especially important because many military personnel live off base in the Tacoma/Pierce County area.

Community Risk Assessment

Threat Assessment. Fort Lewis and McChord AFB ATOs take the lead in threat assessments. Although the ATOs do not personally conduct all types of threat assessments necessarily, the ATOs collect threat assessments that they do not personally conduct or oversee and work with the appropriate functional offices and subject matter experts to ensure that all types of threats are addressed for their respective installations. McChord AFB uses an Air Force OSI template for threat assessments for criminal and terrorist events. CBRNE, toxic industrial chemicals and materials, public health, and natural disaster threats are assessed separately and may follow different models. Fort Lewis assigns numerical values (on a scale of 1–4) to each type of threat to reflect operational capabilities, intentions, and level of activity. Antiterrorism personnel at Fort Lewis referred to JSIVA assessment procedures as a process guide for Fort Lewis threat assessments.

Both Fort Lewis and McChord AFB rely on threat assessments conducted by the local community or Washington state for natural threats, such as volcanic activity or tsunamis. These types of threats are viewed as relatively constant and community-wide, so military personnel focus on assessments for threats that are military-specific or frequently changing.

On the civilian side, both hazard identifications and vulnerability assessments are conducted. These assessments are completed in association with law enforcement, while threat assessments are conducted through a threat early warning and regional intelligence group at Pierce County, colocated with joint tactics, techniques, and procedures, connected to the Washington state joint analytic center. Civilian emergency managers, like their military counterparts, also use statewide threat assessments for natural disasters. Further, at the time of the interviews, the city of Tacoma was developing a business continuity plan in which department directors review historical and perceived threats by geography and region, including flooding and earthquakes, proximity to a military installation or port, and potential for a terrorist event.

The civilian homeland security office uses state-provided guidance to conduct state-mandated hazard identification and vulnerability assessments. Natural and technological hazards are listed and detailed in the state guidance, and the resource also includes worksheets to facilitate hazard identification and risk assessment, vulnerability analysis, mitigation measures, and mitigation planning. For example, the hazard-identification and risk assessment worksheet instructs the analyst to consider hazards in terms of their likelihood of occurrence, locale, impacts, and hazard index.

The VA uses a combination of historical events and other inputs to conduct vulnerability assessments. The emergency manager surveys local community activities and projected events, along with risk factors, such as rail-line locations and major transportation routes, to add to the threat assessment.

Vulnerability Assessment. Vulnerability assessments vary by organization. McChord AFB vulnerability assessments are conducted by individual functional areas (e.g., bioenvironmental engineers, public health, food safety) addressing differing priorities. Once completed, these vulnerability assessments are collected and passed to the ATO.

Fort Lewis personnel referred to JSIVA and the Joint Anti-Terrorism Guide as the process guide for vulnerability assessments and cited forming a tiger team with functional experts for food and other public health vulnerability assessments. Madigan Army Medical Center's Mobilization Disaster Preparedness Committee at Fort Lewis also conducts threat and vulnerability assessments for the hospital. Each department and clinic at Madigan provides input to these assessments, as does the base PHEO. MEDCOM further augments Madigan's health vulnerability assessments.

VA vulnerability assessments are informed by the VA Office of Occupational Safety and Health guidebook; however, the VA emergency manager created his own worksheet to complete the assessments. Using this worksheet, the emergency manager assesses previously identified hazards and their projected impact on human safety and lives, property, and health-care operations.

On the civilian side, the county homeland security office uses state-provided guidance to conduct state-mandated hazard identification and vulnerability assessments. State guidance includes worksheets to facilitate hazard identification and risk assessment, vulnerability analysis, mitigation measures, and mitigation planning. As an example, the vulnerability-analysis worksheet categorizes assets by types (e.g., residential, commercial, hospitals, schools, hazardous facilities) and prompts the analyst to input information on the number of people, number of buildings, and approximate value. Some of the assets reviewed on the civilian side are also analyzed in military vulnerability assessments; however, scope and angle of analysis vary.

Local Planning and Exercises

Tacoma's city manager delegates the majority of responsibility for homeland security to the fire department, public works, and police department, although, by charter, the city manager has responsibility for preparing plans and executing training for emergencies. The fire and police departments develop emergency management plans and first responder training and coordinate these through the city manager's office. Other cities in Pierce County, however, contract with the county for emergency management planning. Communities have planning committees for emergency management, local emergency preparedness coordinating committees (chaired by county health departments), and LEPCs.

Since Washington is a home-rule state (any city of more than 10,000 inhabitants has considerable autonomy), there are fewer state-imposed guidelines, and the counties can do more planning without state involvement. Consequently, Fort Lewis and McChord AFB coordinate with both local cities (i.e., Tacoma and Lakewood) and local agencies, such as Pierce County's Public Health Department and Sheriff's Office.

Generally, military and civilian disaster plans are all-hazards, with functional and event-specific annexes. Fort Lewis and McChord AFB plans are integrated through a top-level contingency plan for the EOC, referencing other emergency management plans developed by functional groups or ESFs (e.g., medical contingency response plan, fire plan, public works plan), which go into greater detail (e.g., checklists). The process by which plans are made and approved differs within the two installations, reflecting, in part, service differences. Given the aforementioned move to joint basing as a result of BRAC, several interviewees suggested that this may change.

Both military and civilian interviewees reported a robust exercise schedule that typically involves several tabletop exercises and at least one major full-scale exercise each year. Civilian interviewees also cited participation in regional or national exercises in addition to exercises focused on Tacoma/Pierce County. Exercises planned by various organizations covered different types of events, such as mass casualty, pandemic influenza, and terrorism. There was some redundancy in exercises, in that more than one organization planned exercises on the same type of event, but, in general, organizations integrated efforts as much as possible, including inviting other organizations to exercises and supporting outside organizations' exercises.

Involvement of the VA and the National Guard is greatly affected by proximity. The VA does not have a significant presence in Tacoma area, as its closest facility is 35 miles away in Seattle. In contrast, Camp Murray, a National Guard installation and the site for Washington's EOC, is close to McChord AFB, and the National Guard's Western Air Defense Sector (responsible for air defense west of the Mississippi) is a tenant unit on McChord AFB. These local National Guard elements are regularly involved in various planning and exercise activities. As a result, informal relationships with National Guard personnel have developed, in addition to official channels of communications. For example, in the event of an emergency, a call can be made to Camp Murray to alert the National Guard commander to alert him or her that a formal request for assistance is making its way through the chain of command.

Key Actors and Coordination in Local Planning

In the Tacoma/Pierce County area, there are many MOUs, inter-service agreements, and coordinating civilian-civilian agreements in place, including fire, police (especially SWAT-team support), and medical. Medical MOAs and MOUs are in place with community facilities as well as with county health departments, in part to develop SNS exercises and coordinate a pandemic influenza response.

Interviews indicated that there may be more military-civilian interaction at the functional level, especially for the medical/public health and firefighting communities. For example, for fire, McChord AFB is integrated into the Pierce County Fire District System and has MAAs in place with University Place, Lakewood, Central Pierce County Fire and Rescue, and others. McChord AFB is especially active within Lakewood, where it responds to a couple of fire incidents per week. In contrast, Tacoma law enforcement has less interaction with its Fort Lewis and McChord AFB counterparts.

Military and civilian organizations have a relatively high level of interaction with respect to emergency management plans and exercises at the functional level. However, it does not appear that Fort Lewis or McChord AFB engages civilian organizations outside of their installations when developing installation-level all-hazards response plans, nor are civilian organizations tasked within installation plans. Interaction does occur between the military and civilian organizations through sharing of plans and verifying contact information. There is also intelligence sharing at the local, state, and federal levels. Further, meetings are held with military and civilian emergency managers to discuss what each organization can contribute in disaster response and to create and execute joint exercises. Lastly, as noted earlier, BRAC has motivated greater interaction between Fort Lewis and McChord AFB personnel, including both those tasked with specific functional responsibilities and those involved in overall emergency management.

Capabilities in the Military and Civilian Sectors

The Tacoma and Pierce County region has a significant number of well-trained people and organized teams to utilize for large-scale incident management. Additionally, both civilians and military personnel agree that the area is very robust in terms of its own capabilities and would require little military assistance in a disaster situation. In the Tacoma/Pierce County area, fire departments have HAZMAT and incident command; police departments have riot response; Tacoma has a SWAT team; and Tacoma is an MMRS city. Further, there are two type-3 incident management teams.

Military assets identified by interviewees as of use to the community include McChord AFB's airfield and hangars, specialized equipment (e.g., the SkyWatch™ tower, which is a bulletproof self-contained unit with elevation capabilities), National Guard mobile brigade sections, Chinook helicopters, C-17s, "fording" vehicles (for use during floods), and explosive-ordnance teams. However, Fort Lewis and McChord AFB report that they would require SWAT-team support from Tacoma, SNS distribution from Pierce County, and the CST from the National Guard. Military installations in the area also tap into community capabilities for support in traffic routing, SAR, and pathways to community-produced information (e.g., Mount St. Helens information and contact information, disease surveillance).

Disaster Preparation and Response Facilitators

Although active duty military personnel rotate frequently through Fort Lewis and McChord AFB, DoD civilian employees, particularly in emergency management functions, help maintain institutional knowledge and often have deep knowledge about the local area (and thus can serve as information sources for new military personnel). These DoD civilians also serve as consistent points of contact for civilian counterparts and remain in the area long enough to develop informal relationships with them.

The Pierce County Emergency Management portal, a locally developed tool, provides a list of every civilian resource available in Pierce County, as well as other information potentially helpful during an incident, such as school blueprints and emergency exits. The portal is also a repository of information about upcoming training events, classes, exercises, and AARs. Fort Lewis and McChord AFB can connect to this portal (and do so, although some reported access problems due to Army information technology [IT] restrictions), but installation personnel have not contributed information to the portal about their own resources.

Disaster Preparation and Response Obstacles

For Fort Lewis and McChord AFB, military regulations create notable hindrances to duties that can be performed by civilian personnel. For instance, regulation dictates who is eligible to fill particular positions (i.e., military or DoD civilian) and the duties to be performed by such positions, and usually these types of personnel are not interchangeable. For example, interviewees at McChord AFB explained that civilian employees cannot do decontamination, nor can they serve on medical-response disaster teams. Further, IT restrictions limit access to potentially useful tools for emergency management, threat assessment, and other activities. These restrictions include limitations on what can be downloaded to installation computers, limited access to the Pierce County Web portal, and lack of wireless communications capabilities. We also were advised that the Army is on a different radio frequency from other emergency management players, and many interviewees on the community side felt that access to military communications systems was limited. Community interviewees further felt that decisionmakers on the military side (e.g., installation commanders) were not willing to accept the NIMS command structure, instead of a military/tactical approach.

On the civilian side, interviewees noted that the need to meet their organization's exercise requirements sometimes prevented them from participating in other entities' exercises, even though they would have liked to participate. There was also discussion of the lead time required for planning major exercise events and advance notification needed to get organizations on board. Lastly, sometimes, plans for joint exercises are canceled due to real-world event priorities.

Another issue for the military is that medical readiness data are lacking for DoD civilian employees and contractors (e.g., immunization status, training level, entitlement to pharmaceuticals), a gap that could hinder disaster response. While this may be true for McChord AFB civilian personnel, the issue was discussed only during interviews with Fort Lewis representatives.

Decision Support Tool: Potential Applications

Many of the suggestions from Tacoma-area interviewees center on ensuring quality of information provided with regard to best practices, intelligence, and the like. Below is a list of characteristics for a decision support tool, as suggested collectively by Tacoma-area interview participants, broken down by phase:

- Planning
  - Create a common operational picture to which all responders contribute.
  - Make it possible to access the documents and checklists of other organizations.
  - Combine and organize response requirements from different organizations so that a comprehensive, prioritized timeline could be constructed.
  - Crosswalk requirements from documents to plans.
  - Identify best practices for integration into capabilities, exercises, and training.
  - Create repository for emergency management plans for all organizations in a region.
  - Prioritize real-time information from different sources and push it out to key local actors.

  - Match actual capabilities to requirement levels.
  - Provide medical readiness data not only for active duty personnel but also for civilians and contractors.
  - On the military side, track training in emergency management planning so that, even if personnel are not currently serving in emergency management jobs, they can be tapped as an emergency management resource.
- Event management
  - Create a common operational picture to which all responders contribute.
  - Show who is doing what, including a checklist that highlights which tasks have been completed and which remain outstanding.
  - Provide portable mapping and plotting programs.
  - Prioritize real-time information from different sources and push it out to key local actors.
    - Example: During an event, hundreds or even thousands of calls can come into a given agency; the tool would be useful if it could automatically organize such calls by type of caller (e.g., first responder, utility worker, private citizen), location, and, in the case of calls containing intelligence, coded for importance or functional relevance.
- Post-event
  - Capture how assets were deployed during the event in order to apply for cost reimbursement.
    - WebEOC creates situational awareness, as well as deployment of resources, logs in order to simplify applying for reimbursement.
  - Encourage focus on future planning across multiple operational periods following an event.
  - Support development and implementation of remediation plans.

Decision Support Tool: Implementation-Related Issues

Interviewees felt that, in order for a tool to be effective, high-level authorities must both endorse the tool and require its use. Further, it should be standardized, ideally at the national level, meet the approval of military IT authorities, as well as be portable (wireless access capabilities). Data sensitivity issues would also have to be worked through and managed (e.g., would civilians be able to see military AARs in their entirety or abbreviated versions?). Lastly, interviewees also indicated that support for training and upkeep of the software would be necessary—that the initial introduction of the software would not be sufficient for implementation.

Summary and Conclusions

The Pierce County and Tacoma area is home to more than 800,000 residents and two major military installations. Overall, interviews with city and county emergency management personnel revealed that the area is well equipped to respond to disasters and has many individuals who have extensive training and experience in emergency management. Tacoma delegates the majority of emergency management to the different functional agencies (e.g., police, fire, public works), whereas other cities in Pierce County contract with the county for emergency management planning. As a result, disaster plans are typically all-hazards with functional and event-specific annexes. This is also true at Fort Lewis and McChord AFB, where functional plans are integrated through a top-level contingency plan for the EOC.

Civilian-civilian coordination in planning in the area seems extensive, with MOUs and inter-service agreements between fire, police, and medical functions. Military-military coordination exists, although primarily for medical support. Civilian-military coordination is strong between McChord AFB and civilian fire departments, but interviews did not reveal other similar coordination for other functions, such as law enforcement. Military and civilian organizations do share intelligence, however, and military-civilian meetings have been held to discuss disaster response capabilities and to plan and execute joint exercises. Also, Fort Lewis and McChord personnel tasked with emergency management responsibilities acknowledge that a disaster that affects Tacoma or Pierce County will likely also affect the installations.

City of Las Vegas and Clark County, Nevada (September 2008)

Background

About 80 percent of Nevada is federally owned and managed, including land managed by DOE and DoD. Las Vegas and Clark County, the county in which the city is located, are geographically isolated from other metropolitan areas. Local emergency management officials report that they are very self-reliant and plan based on the assumption that they will be the first and only responders for at least four days following any major disaster. Further, the city of Las Vegas and Clark County have a unique relationship, in part due to geography typified by uniquely shaped boundaries that do not make distinctions between military and the surrounding communities clear. McCarran International Airport and the Las Vegas Strip are both situated on county land, and Clark County directs emergency management within the geographic region, including forging relationships with Nellis AFB. Interviewees indicated that the county also controls a larger proportion of regional and municipal services (e.g., local social services) than is typical for a county.

The city of Las Vegas is home to a Fusion Center (see definition in Appendix A and description in Appendix B), which serves as an intelligence clearinghouse for a variety of local, state, and federal actors; the police department maintains the All-Hazards Regional Multi-Agency Operations and Response team, which is an integrated HAZMAT and bomb-squad unit that was developed with support from the National Guard CST.

The leading hotels and casinos in Las Vegas are like mini-cities, each with its own set of resources (e.g., power supply, large kitchens, housing, and security departments). These organizations do not tend to participate proactively in emergency planning but are quick to contact the local government in the event of a disaster to offer assistance. According to civilian emergency management personnel, there are about 7,000 security personnel working for properties on the Las Vegas Strip.

Nellis AFB is home to more air squadrons than any other U.S. Air Force base and serves as a major training facility for both U.S. and foreign military aircrews. Nellis is the only major military installation in the Las Vegas/Clark County area and manages a significant area of land not adjacent to the installation itself. Nellis AFB encompasses more than 14,000 acres, and almost 8,100 active duty personnel are assigned to it (see Table 4.1 in Chapter Four).

Neither the city of Las Vegas nor Clark County has experienced any major disaster in recent years. The most severe event was flooding in 1990. In addition, surges of tourists into Las Vegas occur predictably. For example, Las Vegas and Clark County regularly plan for the New Year's Eve holiday on the Las Vegas Strip, a gathering of people estimated to be second only to the one at New York City's Times Square in size and scope. Occasionally, there are large events, such as the NBA All-Star Game, which pose some difficulties related to law enforcement. Other recent events with potential but avoided disastrous consequences include the purposeful release of veterinary-grade anthrax and an attempt to release ricin. Further, there was a fire at the Monte Carlo hotel and a near-miss event with a runaway chlorine tanker, which could have led to disastrous consequences.

Terrorism (both domestic and international) is always a high concern in the area, given its assets and the influx of tourists. Other potential hazards to the region include communicable disease outbreaks, wildfire, flash flooding, earthquake, river flooding, and drought. There also concern in particular regarding low medical capacity—indeed, Nevada ranks 49th in the nation for numbers of hospital beds.

Interview Profile

As shown in Table 4.2 in Chapter Four, the RAND team conducted ten interviews in the Las Vegas area: four with representatives from Nellis AFB, five with representatives from the city of Las Vegas and Clark County, and one with a representative from the VA. At Nellis AFB, interviewees worked in emergency management, CBRNE, civil engineering, medicine, bio-environmental engineering, firefighting, public health, public safety, and antiterrorism. Interviews with officials from the city of Las Vegas and Clark County included those working in emergency management, public health, public safety, antiterrorism, firefighting, and medical response. A VA official was also interviewed about the VA role in emergency management. We were unable to interview a representative from the executive branch of government.

Definitions of Community

Most Nellis AFB personnel expressed a narrow view of community, focusing primarily on those areas for which they were directly responsible. For example, Nellis AFB is responsible for nonadjacent facilities that include Creech AFB, 45 miles northwest of the base; Nevada Test and Training Range, 3.1 million acres used for various testing operations across multiple counties; and Tonopah Test Range, conducting aeronautical research and development.

However, interviewees from the city of Las Vegas and Clark County often defined the community as Clark County, given the presence of unincorporated areas, the strong role of the county as service provider, and the geographic isolation of the area. The VA area of responsibility includes southern Nevada but focuses mostly on Clark County. Both military and civilian interviewees recognized that many events that would affect Las Vegas would also affect Nellis AFB and vice versa, since Nellis AFB is only eight miles away from the center of downtown Las Vegas.

Overall, the broadest views of community among Nellis AFB personnel were expressed by individuals responsible for public health, security forces, or intelligence gathering or analysis. Military public health personnel work regularly with their civilian counterparts on epidemiological matters and plans for mass prophylaxis, including delivery at points of distribution. Interviewees with security responsibilities noted that, when a military asset is involved in an event off base, military authorities can create a National Defense Area in order to temporarily own the jurisdiction for that disaster site and maintain control of sensitive information and equipment. Interviewees with intelligence responsibilities noted their regular participation in local terrorism planning with civilian counterparts.

Community Risk Assessment

In general, the city of Las Vegas and Clark County use threat assessments and hazard vulnerability assessments to inform their plans. Las Vegas and Clark County plans are typically reviewed and revised as needed on an annual basis, although many remain largely unchanged from when first drafted. Revisions typically involve updating key contact information, incorporating new capabilities, or responding to a specific requirement (e.g., a pandemic influenza plan with a communications annex was a CDC requirement that had to be added). Some plans are shared between Clark County and Nellis; however, it is often between functional specialties, and coordination is not transparent to the larger organizations.

Threat Assessment. Nellis AFB coordinates the installation's threat assessment through its antiterrorism office, which, in turn, works with the integrated local Fusion Center to identify local terrorist threats to the installation. Threats related to toxic industrial chemicals or materials, food, water, and natural disasters are incorporated into the assessment as well. Subject matter experts are tapped as need to inform the assessments for these specific, non–terrorism-related threats. Nellis AFB's antiterrorism office also coordinates monthly reviews of the threat assessments with Air Force OSI, the FBI, the National Guard, and local law enforcement agencies to identify any necessary revisions—especially to the top local threat. Nellis AFB medical (hospital) staff contribute to the overall installation threat assessment and separately handle smaller-scale comprehensive assessments related to epidemic threats.

The city of Las Vegas and Clark County conduct threat assessments locally, based primarily on history. Stakeholders coordinate to consider the area's hazards and have agreed on five major threats, based on history (i.e., probability of threat) as well as potential impact (i.e., vulnerabilities and consequences). These assessments are informed by data from the local Fusion Center as well as information from CDC. Community threat assessments are not as formal as those done at Nellis AFB, nor are they done with the same consistency, occurring annually or less frequently.

Military-civilian interactions related to threat assessment consist primarily of intelligence sharing and representation on each other's committees (e.g., the military TWG and the civilian LEPC) to discuss threats. Civilian participation in Nellis emergency planning is limited by the classified nature of some of Nellis's planning information.

Vulnerability Assessment. Nellis AFB conducts a vulnerability assessment annually, informed by its threat assessment. Nellis follows the Air Force vulnerability-assessment process, and all key military entities contribute to the assessment in some way (e.g., OSI, special forces, intelligence, medical readiness, communications squadron, explosive ordnance, fire). They use such tools as the M-SHARP (which assesses vulnerabilities to targets on the basis of their mission, symbolism, history, accessibility, recognizability, population, and proximity) and CARVER (which likewise assesses targets on the basis of their criticality, accessibility, recoverability, vulnerability, effect, and recognizability), but these tools do not replace subject matter expertise and intuition. The focus is mostly on the base itself and on the nonadjacent military properties for which the base is responsible.

Vulnerabilities are split up by functional area, with bioengineering focusing on water- and air-related hazards; public health addressing biological hazards; and emergency managers examining natural disasters, CBRNE, and WMD. The hospital does a separate vulnerability assessment. This assessment includes the VA, since Nellis and the VA currently share the hospital facility. Nellis is starting to rely more on civilian information because the installation's employees live off base and thus become part of the civilian planning population in terms of communications, water, and power during an emergency situation. For example, Nellis leadership sends toxic industrial chemicals and materials and food and water assessments and information about water and fuel supplies to civilian counterparts, but they cannot share much of the base's own vulnerability assessment because of classification issues.

Overall, vulnerability assessments appear to be lacking for the city of Las Vegas and Clark County, except for public health, which is conducted at the county level and based on CDC guidance. The Las Vegas Metropolitan Police Department (LVMPD) had contracted out a vulnerability assessment; however, the contractor is reported to have fallen short on its responsibilities. The contracted project, Silver Shield, only identified tier I and tier II critical infrastructure, did not include any community-level stakeholders, and reportedly did not include an analytic component. Community stakeholders did indicate, however, that the LEPC helps identify critical infrastructure, and they reported that mitigation strategies have been developed to help protect many critical local assets.

The VA does not have a traditional requirement for its hazard vulnerability assessment in Las Vegas because of its unusual status: sharing the VA hospital with and leasing administrative offices from Nellis AFB. When its new hospital is built, the VA will use the VA Emergency Management Program Guidebook (2002), which is very detailed. At the time of the interview, VA did an annual vulnerability assessment, guided by MOFH's emergency plans, with its emergency planning committee, which informs its emergency management plan and its continuity-of-operations readiness plan. Scenarios are chosen by the committee members' estimation of what is most likely to occur in their area.

Local Planning and Exercises

Both military and civilian interviewees described their local response plans as all-hazards. The military developed plans with functional annexes, rather than hazard- or event-specific annexes. In some cases of high-level plans, entities responsible for specific functions (e.g. fire, EMS, medical) may develop their own checklists on how to execute their specific responsibilities. Similarly, civilian leaders created all-hazards plans with roles and responsibilities by function.

Disaster planning is largely undertaken separately by civilian and military entities; however, exercises are more likely to be joint military-civilian. In the past, the VA participated in exercises with Nellis AFB. Large local events, such as the Nellis air show or NASCAR races, may involve more joint planning for the specific event. Nellis AFB and Clark County do share their plans with each another, but the degree to which lower-level functional counterparts know one another is inconsistent.

Overall, planning templates vary. Civilians sometimes use FEMA guidance, the Homeland Security Emergency Plan, or CDC recommendations (in the case of public health). The LEPC also uses an emergency planning template from Texas.

Local civilian emergency response plans are drafted with stakeholder input and through coordination with stakeholders, and then tested in exercises. Plans are revised in accordance with lessons learned during these exercises. Conversely, at Nellis AFB, individual units take base or Air Force plans and create unit-based checklists to support the plans.

Draft MAAs for public health and medical services acknowledge both civilian and military exercise requirements and encourage community-wide exercises as much as possible (e.g., at least one off-base community-wide operational exercise—not a tabletop exercise); AFI 10-2501 requires a mass casualty exercise at least annually, and AFI 41-106 has a similar requirement for a CBRNE exercise involving off-installation responders. Exercises appear to be ongoing, ranging from tabletop to full-scale operational exercises, and requirements are typically defined by differing agencies and grantmakers. Nellis AFB reportedly participates in 50 or more exercises each year, including tabletop and functional exercises and drills; several civilian agencies report exercising at least monthly. Exercises are typically situation specific (e.g., Hoover Dam break or a workplace hostage scenario), although a mass casualty component is standard. There are not as many multifunctional exercises, especially military-civilian, as there are function-specific scenarios and exercises. The VA conducted exercises with Nellis AFB in the past; however, the VA had not done so recently at the time of the interviews, nor did it have any formal assistance agreements. The VA does conduct a nationwide pandemic influenza tabletop exercise each year.

Key Actors and Coordination in Local Planning

Interviewees reported that the Clark County LEPC is an important hub for interactions both across local civilian agencies (and the private sector) and with Nellis AFB, which is a member of the committee and has a seat in the Clark County EOC. Nellis AFB's ties are mostly with emergency management personnel at the county level through the LEPC and less so with City of Las Vegas emergency management personnel. There were also strong relationships with public health.

The MOFH, a joint VA-Air Force hospital located at Nellis AFB, serves as a basis for limited VA interaction with the civilian medical community—e.g., as a member of the local health and safety officers group (as is Clark County Public Health) and the infectious disease committee.

The National Guard assists civilian stakeholders through CST efforts, as well as augmenting law enforcement personnel for the surge in tourists to Las Vegas each New Year's Eve. This appears to be the extent of interaction with the Guard.

Capabilities in the Military and Civilian Sectors

Interviewees felt that local civilian and military agencies in Clark County have unique capabilities that can complement one another. For example, civilian capabilities include a SWAT team, medical trauma center, and CBRNE support. Unique or important Nellis capabilities include explosive-ordnance disposal, radiological monitoring equipment, and watchtowers and other security-related equipment that support civilian security each New Year's Eve. These assets can be shared from civilian to military or vice versa.

Interviewees view exercises as a key way for agencies to learn about one another's capabilities; this is especially true with respect to the community's learning about Nellis's capabilities. Generally, the city of Las Vegas and Clark County appear to have more capabilities to offer Nellis AFB than vice versa. Civilian authorities indicated that they are not familiar with Nellis AFB capabilities and do not know how to request or mobilize them. This can be attributed to communities having more stable resources and fewer legal constraints on usage. However, a draft local MAA provided to the RAND study team in September 2008 for public health and medical services outlines the type of support and process for requesting support from the MOFH.

Clark County and the LVMPD use the DHS-recommended free IRIS, which is an all-hazards resource inventory for the availability of resources for mutual aid. The LVMPD needs only to input its requests.

Disaster Preparation and Response Facilitators

Several national-level initiatives drive collaborative local emergency planning in the Clark County area. First, the national emergency response organization and processes promulgated by DHS through NIMS and its ICS have reportedly helped bridge differences across local agencies, such as through common language and command structure. However, interviewees indicated that the Air Force NIMS and community NIMS still have notable differences, such as how a particular NRF ESF is used. Local agencies have begun to use "plain talk" as a strategy to overcome remaining barriers created by differences in language and terms within the emergency response context.

Second, grants, such as those from DHS and HHS, drive collaboration across civilian agencies—for example, from the HHS Office of the ASPR, pandemic influenza planning, MMRS, and UASI grant programs. Finally, the Fusion Center brings together military (both Nellis AFB security forces personnel and Air Force OSI agents), the local community, and state and federal actors to share locally relevant intelligence.

Communications interoperability is another facilitator of local preparedness planning. Interviewees reported that system interoperability is good but suffers from some limitations: Civil engineering and the National Guard CST both have ACU-1000 equipment (a Raytheon product that serves as a universal translator that can connect dissimilar radio systems), but this is the extent of communications interoperability in the area. Common communications do take place among LEPC members via other means, however: The committee relies on a wiki (computer-based collaborative tool) for developing and refining Clark County plans. Several interviewees commented that this is very effective.

Disaster Preparation and Response Obstacles

Obstacles to local coordination of emergency preparedness planning in the Clark County area range from conceptual to practical: from fundamental legal differences between military and civilian actors that prohibit engagement to differences in communications technologies and terms and the planning tools these different actors use.

A first perceived barrier is legal. Posse Comitatus limits use of Nellis AFB security forces for law enforcement off base. Some interviewees emphasized that Nellis AFB should always be the player or resource of last resort for civilian response, noting the requirement that civilians exhaust all for-profit service providers before the federal government steps in and provides services that must subsequently be reimbursed by the state.

In the event of a terrorist attack, some civilian interview participants explained that military bases may be locked down and that civilians thus should not expect to receive any support. However, several Nellis AFB interviewees noted that a terrorist attack would be an unusual situation in terms of requiring an installation lockdown and that, instead, in many disaster scenarios, military assets would be available. Further, Nellis AFB personnel were aware that civilians will not provide assistance to a CBRNE event on the installation: Civilian ambulances will not transport contaminated patients, and local hospitals will not accept contaminated vehicles.

Personnel turnover poses another difficulty, particularly for Nellis AFB and local civilian hospitals. High turnover rates, especially in active duty military personnel due to permanent changes of station and deployments, make it difficult to know the right person to contact in order to request support. In addition, hospitals find it especially difficult to retain people who are trained to use all of the communications technologies needed for emergency response. One strategy employed to overcome this obstacle is to rely more on civilian employees at Nellis AFB, especially in civil engineering and exercises, evaluation, and training organizations.

A particular obstacle on the civilian side is the lack of security clearances and the lack of reciprocity of security clearances when dealing with state, federal, and military actors. To circumvent this issue, the Las Vegas Fusion Center focuses on unclassified data, particularly local data that analysts in Washington, D.C., might not readily have. The goal is to complement top-down information that may be classified with local intelligence that comes from the bottom up. Even civilian-to-civilian information sharing can be a problem, however: Law enforcement does a classified assessment of facilities every four years, but some civilian interviewees reported having limited, if any, access to these findings.

Despite some positive elements described in the preceding section, communications remain an obstacle to coordinated local planning in the Clark County area. There are fundamental technology differences between military and local actors and even across civilian actors (e.g., use of different radio frequencies), different call signs and codes, and even different dialogue, especially when dealing with local, federal, and military responders. Technology-based crosswalks, such as Raytheon's aforementioned interconnectivity system, the ACU-1000, help somewhat, as do "communications rodeos" (agencies coming together to develop communications solutions to unexpected problems that emerge in various scenarios). But beyond incompatibilities in communications technologies are differences in organizational cultures. For example, when civilians ask for a capability, they mean in minutes or hours, whereas Nellis AFB personnel may think more in terms of days.

Finally, WebEOC is commonly used by community stakeholders, but there are different versions of the package—city, county, and state each have their own version. The LVMPD in particular mentioned not being allowed to use the county's version anymore because it did not adhere to county guidance regarding WebEOC use. Clark County noted that Nellis AFB has the ability to log into the county's WebEOC, although some functional entities at Nellis AFB did not seem not aware of this.

Decision Support Tool: Potential Applications

Interviewees described the need for decision support tools with the following characteristics, broken down by phase:

- Planning
  - asset visibility, especially for local assets and accessible Nellis AFB resources
  - links and references to best practices (similar to those in inspector-general reports)
  - automated analysis of AARs and other data from exercises in order to reveal systematic problems across exercises
- Event management
  - real-time asset status and availability information for SWAT teams and the like (requested by military interviewees)
  - situational awareness
  - real-time chat capabilities
  - mimicking of the incident mapping function of the Air Force's Theater Battle Management Core System
  - an electronic tactical worksheet, complete with plans for an organizational chart and assignments (requested by military interviewees)

- accurate information source for Public Affairs
- checklists outlining DSCA and how to request resources from the military, state, federal, and other entities in an effort to streamline the bureaucratic process
- informed analysis for decisionmakers enacting policy or requesting state or federal support

- Post-event
  - ability to download actions into AARs
  - capture during-event dialogue for official record and to feed into AARs (similar to the Theater Battle Management Core System)
  - repository for AARs and other evaluations of exercises and events.

Decision Support Tool: Implementation-Related Issues

Some of the interviewees felt that too many computer-based tools already existed. Others were resistant to using a new and unproven tool and preferred to rely on proven tools with which they had experience. At the time of the interviews, there were also many key local actors who relied on paper-based systems, such as field manuals and NIMS boards, partially due to habit and partially because they felt that technology was not always reliable.

Interviewees indicated that they would find it very difficult to envisage a tool that could be used by military and civilian responders during an event, due to classification issues, uncertain access to wireless Internet, and the possibility or even likelihood of computers or Internet access going down or being overloaded during an event. Participants also noted the difficulty of using any tool that required manual input of data, especially at their staffing levels, albeit recognizing that user input would probably be necessary given how many disparate systems are currently used. One suggestion offered by the LEPC were wikis, which, as noted previously, were used by LEPC members to log their plan revisions, a process that was deemed very effective.

Other suggestions from participants revealed the utility of a scalable tool, one that could work on smaller scales as well as be built up. Interviewees also thought it would be useful if the tool could "plug and play"—i.e., be interoperable with all other operating systems currently being used (e.g., WebEOC, IRIS). Lastly, some respondents said that the most useful contribution would be a dynamic information-sharing tool, with as little time lag as possible.

Summary and Conclusions

Las Vegas and Clark County are unique areas due to their atypical boundaries and the federal status of much of Nevada's land. Nellis AFB is also noteworthy as the major training facility for military aircrews. The presence of the Fusion Center serves as an intelligence clearinghouse and brings together military and civilian personnel.

At the time of the interviews, both military and civilian emergency management were conducting regular exercises and planning. The VA in the area sends its emergency plans to Nellis AFB for incorporation, as it has a unique relationship with Nellis, while the VA shares MOFH. Both Nellis and the VA conduct vulnerability assessments; the city of Las Vegas and Clark County do not conduct such assessments, except in the case of public health hazards. The Clark Country LEPC brings together representatives from the city, county, and Nellis AFB, providing an opportunity for coordination in disaster planning that might not otherwise occur.

APPENDIX F

Social Network Analysis Survey Protocol

Mapping Emergency Response Networks

In order to develop a general network for disaster preparedness and response for each community, respondents were given these instructions and were asked to fill in the table in the following survey. While not limited to the organizations listed, suggestions of potential network members were provided on p. 9 of the survey as a guideline. Participants were also instructed to attach additional pages if they wished to list more than 12 network members.

Military-Community Readiness for Major Disaster Events: Optimizing Risk-Based Capability Planning

Center for Military Health Policy Research
A JOINT ENDEAVOR OF RAND HEALTH AND THE RAND NATIONAL DEFENSE RESEARCH INSTITUTE

Study Overview and Nature of Participation

Study Description

RAND Corporation, a non-profit research institution (www.rand.org) that includes federally funded research and development centers for the Department of Defense, is conducting a research project at the request of the Assistant Secretary of Defense (Health Affairs). RAND has been asked to facilitate local readiness planning for major disasters by developing a decision support tool that can help communities to:

1. Assess risks
2. Identify capabilities
3. Coordinate planning for major disasters across local stakeholders including the Department of Defense, local government, and local Veterans Administration (VA) health providers.

The focus of this research effort is on communities that contain a military installation within their vicinity.

The final product of this research will be an input-output decision support template that can be used by both military and civilian emergency planners. It will be designed to take advantage of current data on requirements and capabilities at federal, state, and local levels, as well as be tailored to local circumstances.

To ensure that the design of this template is fully informed by both military and civilian planners, the RAND research team is interviewing representatives from military installations, civilian response planners in neighboring cities, and local VA health providers to understand better how such planning occurs, how resources are identified and shared, and who is involved in preparing to respond to emergency situations. The team also will solicit ideas about a potential decision support template. In addition, the RAND research team is administering a survey, which is intended to complement the interviews by helping to establish a picture of the all-hazard response network present within the local community. Individuals that participate in RAND interviews will be asked to complete this survey, along with additional individuals knowledgeable of their local community's all-hazard response network.

Survey Participation

You have been asked to participate in this survey because of your professional knowledge of the all-hazards response network in your local community. In this survey, you will identify organizations that you or your organization interact with in the context of disaster preparedness and then describe those interactions through a series of multiple choice questions. Depending on how many organizations you list, the survey should take you about 20 to 30 minutes to complete. Your participation in this survey is entirely voluntary. You may decline to participate, and you are free to skip any questions that you prefer not to answer. In addition, although we request that you provide your name, title, and organization in conjunction with your survey response, we will not associate any individually identifiable information with your survey responses. Instead, a code will be assigned that links your survey to you indirectly.

The results of this survey, along with those stemming from RAND's interviews and document review, will inform the development of the aforementioned decision support tool. In addition, findings based on the survey will be included in a report prepared for the research sponsor. We may discuss individual observations as well as patterns across our surveys, but we will not cite your name or title. We may attribute findings by organization (for example, [city name] emergency management office, [city name] police department, or installation management function at [installation name]).

After the study is complete, all information linking you to this survey will be destroyed. A copy of your survey that does not contain your name or other personal identifying information will be retained to inform future research on this topic.

If you have questions about this survey or would like to be notified when the final report is publicly released, please contact Ms. Darcy Noricks at dnoricks@rand.org or 310-393-0411 x6273.

In order for us to be able to include your survey in our analysis, we need to know your name, title and organization. (As a reminder, all information linking you to this survey will be destroyed after the study is complete.) Please fill out the following:

Printed Name ____

Title ____

Organization ____

MAJOR DISASTER RESPONSE NETWORK

We are interested in what the general network for disaster preparedness and response looks like in your community. This will help us to know a little more about the coordination among various community institutions. With that in mind, we'd like your help in thinking about a major disaster response network for your local community.

KEY TERM: The major disaster preparedness and response network includes preparation and response to natural (e.g., public health emergencies, natural disasters) or man-made events (e.g., HAZMAT, terrorism) that have the potential for catastrophic consequences.

Please fill out the table below. To do so, please follow these steps:

STEP 1: In the first column, please list the local organizations with whom you or your office interact as part of your community's major disaster response network and which you consider MOST IMPORTANT to your network. They may include, but are not limited to, organizations related to health, emergency services, physical infrastructure, community organizations, businesses, and military offices (for examples, see back page).

> **Note for Military Respondents:** After listing your key community-based contacts, please list installation-based contacts with whom you or your office interact that you consider MOST IMPORTANT to your major disaster preparedness and response network. Limit these installation-based contacts to organizations who perform functions different from your office (e.g., communications, security, medical) and with whom you have interacted within the past three years.

STEP 2: In the second column, please indicate the planning dimension for which that organization is responsible (check all that apply). If you check the box marked "Other" please identify a planning dimension, in writing, below the "Other" box.

STEP 3: In the third column, please indicate the function that organization serves in the major disaster response network (check all that apply):

a) Communications: Includes risk communication to the public, between people & organizations, and media relations.
b) Individual Assistance: Includes assistance intended to meet immediate needs like food, housing, hygiene care, childcare, clothing, transportation.
c) Health/Medical: Includes all health care and medical needs including mortuary, stockpile, and mass care.
d) Continuity of Services: Includes continuity of services such as public assistance checks, social security administration, utility services, public transit systems, etc.
e) Security/Public Safety: Includes services to provide law and order to the disaster area.

STEP 4: Next, place a checkmark in the box in column 4 to best describe the type of agreement (e.g. mutual aid agreement) you have with the organization with regard to preparing/responding to a major disaster (check all that apply). If you check more than one type of agreement, please note any rationale behind the multiple agreements in the space beneath each agreement type.

STEP 5: Next, place a checkmark in the box in column 5 which best describes your planning activities (exercise/drills and joint planning) as part of your major disaster preparedness activities.

STEP 6: Finally, place a checkmark in the box in column 6 which best describes the frequency of your contact with this organization in the context of your disaster preparedness activities.

We appreciate your participation in this survey. We will pick it up from you in person during our interview, or we will provide you with a self-addressed envelope to return it to us.

(1) NAME OF ORGANIZATION (Enter name of organization in each box)	**(2) DIMENSION OF PLANNING** (Check all that apply)	**(3) FUNCTION** (Check all that apply)	**(4) AGREEMENT** (Check all that apply)	**(5) TYPE OF PLANNING** (Check all that apply)	**(6) FREQUENCY OF INTERACTION** (Check only one)
#1 –	☐ Natural Disaster ☐ Public Health Emergency ☐ Manmade Hazard (e.g., HAZMAT) ☐ Terrorism ☐ Other (Describe)	☐ Communications ☐ Individual Assistance ☐ Joint Planning ☐ Continuity of Services ☐ Security/Public Safety	☐ Memorandum of Understanding/Agreement ☐ Contract for Services ☐ Informal Agreement ☐ No Agreement	☐ Joint Planning Together ☐ No Plans to Exercise/ Drill Together ☐ Planning to Exercise/ Drill Together in Future ☐ Previously Exercised/ Drilled Together	☐ Rarely Interact ☐ Sometimes Interact ☐ Frequently Interact
#2 –	☐ Natural Disaster ☐ Public Health Emergency ☐ Manmade Hazard (e.g., HAZMAT) ☐ Terrorism ☐ Other (Describe)	☐ Communications ☐ Individual Assistance ☐ Joint Planning ☐ Continuity of Services ☐ Security/Public Safety	☐ Memorandum of Understanding/Agreement ☐ Contract for Services ☐ Informal Agreement ☐ No Agreement	☐ Joint Planning Together ☐ No Plans to Exercise/ Drill Together ☐ Planning to Exercise/ Drill Together in Future ☐ Previously Exercised/ Drilled Together	☐ Rarely Interact ☐ Sometimes Interact ☐ Frequently Interact
#3 –	☐ Natural Disaster ☐ Public Health Emergency ☐ Manmade Hazard (e.g., HAZMAT) ☐ Terrorism ☐ Other (Describe)	☐ Communications ☐ Individual Assistance ☐ Joint Planning ☐ Continuity of Services ☐ Security/Public Safety	☐ Memorandum of Understanding/Agreement ☐ Contract for Services ☐ Informal Agreement ☐ No Agreement	☐ Joint Planning Together ☐ No Plans to Exercise/ Drill Together ☐ Planning to Exercise/ Drill Together in Future ☐ Previously Exercised/ Drilled Together	☐ Rarely Interact ☐ Sometimes Interact ☐ Frequently Interact

(1) NAME OF ORGANIZATION (Enter name of organization in each box)	**(2) DIMENSION OF PLANNING** (Check all that apply)	**(3) FUNCTION** (Check all that apply)	**(4) AGREEMENT** (Check all that apply)	**(5) TYPE OF PLANNING** (Check all that apply)	**(6) FREQUENCY OF INTERACTION** (Check only one)
#4 –	☐ Natural Disaster ☐ Public Health Emergency ☐ Manmade Hazard (e.g., HAZMAT) ☐ Terrorism ☐ Other (Describe)	☐ Communications ☐ Individual Assistance ☐ Joint Planning ☐ Continuity of Services ☐ Security/Public Safety	☐ Memorandum of Understanding/Agreement ☐ Contract for Services ☐ Informal Agreement ☐ No Agreement	☐ Joint Planning Together ☐ No Plans to Exercise/ Drill Together ☐ Planning to Exercise/ Drill Together in Future ☐ Previously Exercised/ Drilled Together	☐ Rarely Interact ☐ Sometimes Interact ☐ Frequently Interact
#5 –	☐ Natural Disaster ☐ Public Health Emergency ☐ Manmade Hazard (e.g., HAZMAT) ☐ Terrorism ☐ Other (Describe)	☐ Communications ☐ Individual Assistance ☐ Joint Planning ☐ Continuity of Services ☐ Security/Public Safety	☐ Memorandum of Understanding/Agreement ☐ Contract for Services ☐ Informal Agreement ☐ No Agreement	☐ Joint Planning Together ☐ No Plans to Exercise/ Drill Together ☐ Planning to Exercise/ Drill Together in Future ☐ Previously Exercised/ Drilled Together	☐ Rarely Interact ☐ Sometimes Interact ☐ Frequently Interact
#6 –	☐ Natural Disaster ☐ Public Health Emergency ☐ Manmade Hazard (e.g., HAZMAT) ☐ Terrorism ☐ Other (Describe)	☐ Communications ☐ Individual Assistance ☐ Joint Planning ☐ Continuity of Services ☐ Security/Public Safety	☐ Memorandum of Understanding/Agreement ☐ Contract for Services ☐ Informal Agreement ☐ No Agreement	☐ Joint Planning Together ☐ No Plans to Exercise/ Drill Together ☐ Planning to Exercise/ Drill Together in Future ☐ Previously Exercised/ Drilled Together	☐ Rarely Interact ☐ Sometimes Interact ☐ Frequently Interact

(1) NAME OF ORGANIZATION (Enter name of organization in each box)	**(2) DIMENSION OF PLANNING** (Check all that apply)	**(3) FUNCTION** (Check all that apply)	**(4) AGREEMENT** (Check all that apply)	**(5) TYPE OF PLANNING** (Check all that apply)	**(6) FREQUENCY OF INTERACTION** (Check only one)
#7 –	☐ Natural Disaster ☐ Public Health Emergency ☐ Manmade Hazard (e.g., HAZMAT) ☐ Terrorism ☐ Other (Describe)	☐ Communications ☐ Individual Assistance ☐ Joint Planning ☐ Continuity of Services ☐ Security/Public Safety	☐ Memorandum of Understanding/Agreement ☐ Contract for Services ☐ Informal Agreement ☐ No Agreement	☐ Joint Planning Together ☐ No Plans to Exercise/ Drill Together ☐ Planning to Exercise/ Drill Together in Future ☐ Previously Exercised/ Drilled Together	☐ Rarely Interact ☐ Sometimes Interact ☐ Frequently Interact
#8 –	☐ Natural Disaster ☐ Public Health Emergency ☐ Manmade Hazard (e.g., HAZMAT) ☐ Terrorism ☐ Other (Describe)	☐ Communications ☐ Individual Assistance ☐ Joint Planning ☐ Continuity of Services ☐ Security/Public Safety	☐ Memorandum of Understanding/Agreement ☐ Contract for Services ☐ Informal Agreement ☐ No Agreement	☐ Joint Planning Together ☐ No Plans to Exercise/ Drill Together ☐ Planning to Exercise/ Drill Together in Future ☐ Previously Exercised/ Drilled Together	☐ Rarely Interact ☐ Sometimes Interact ☐ Frequently Interact
#9 –	☐ Natural Disaster ☐ Public Health Emergency ☐ Manmade Hazard (e.g., HAZMAT) ☐ Terrorism ☐ Other (Describe)	☐ Communications ☐ Individual Assistance ☐ Joint Planning ☐ Continuity of Services ☐ Security/Public Safety	☐ Memorandum of Understanding/Agreement ☐ Contract for Services ☐ Informal Agreement ☐ No Agreement	☐ Joint Planning Together ☐ No Plans to Exercise/ Drill Together ☐ Planning to Exercise/ Drill Together in Future ☐ Previously Exercised/ Drilled Together	☐ Rarely Interact ☐ Sometimes Interact ☐ Frequently Interact

Page 7 provides space for you to list organizations 10-12; if you have more than 12 organizations, photocopy page 7 and attach to the back of this instrument.

If you would like to list more than 12 organizations, please photocopy this page before beginning and attach to the back of this instrument.

(1) NAME OF ORGANIZATION (Enter name of organization in each box)	**(2) DIMENSION OF PLANNING** (Check all that apply)	**(3) FUNCTION** (Check all that apply)	**(4) AGREEMENT** (Check all that apply)	**(5) TYPE OF PLANNING** (Check all that apply)	**(6) FREQUENCY OF INTERACTION** (Check only one)
#10 –	☐ Natural Disaster ☐ Public Health Emergency ☐ Manmade Hazard (e.g., HAZMAT) ☐ Terrorism ☐ Other (Describe)	☐ Communications ☐ Individual Assistance ☐ Joint Planning ☐ Continuity of Services ☐ Security/Public Safety	☐ Memorandum of Understanding/Agreement ☐ Contract for Services ☐ Informal Agreement ☐ No Agreement	☐ Joint Planning Together ☐ No Plans to Exercise/ Drill Together ☐ Planning to Exercise/ Drill Together in Future ☐ Previously Exercised/ Drilled Together	☐ Rarely Interact ☐ Sometimes Interact ☐ Frequently Interact
#11 –	☐ Natural Disaster ☐ Public Health Emergency ☐ Manmade Hazard (e.g., HAZMAT) ☐ Terrorism ☐ Other (Describe)	☐ Communications ☐ Individual Assistance ☐ Joint Planning ☐ Continuity of Services ☐ Security/Public Safety	☐ Memorandum of Understanding/Agreement ☐ Contract for Services ☐ Informal Agreement ☐ No Agreement	☐ Joint Planning Together ☐ No Plans to Exercise/ Drill Together ☐ Planning to Exercise/ Drill Together in Future ☐ Previously Exercised/ Drilled Together	☐ Rarely Interact ☐ Sometimes Interact ☐ Frequently Interact
#12 –	☐ Natural Disaster ☐ Public Health Emergency ☐ Manmade Hazard (e.g., HAZMAT) ☐ Terrorism ☐ Other (Describe)	☐ Communications ☐ Individual Assistance ☐ Joint Planning ☐ Continuity of Services ☐ Security/Public Safety	☐ Memorandum of Understanding/Agreement ☐ Contract for Services ☐ Informal Agreement ☐ No Agreement	☐ Joint Planning Together ☐ No Plans to Exercise/ Drill Together ☐ Planning to Exercise/ Drill Together in Future ☐ Previously Exercised/ Drilled Together	☐ Rarely Interact ☐ Sometimes Interact ☐ Frequently Interact

If you would like to list more than 12 organizations, please photocopy this page before beginning and attach to the back of this instrument.

POSSIBLE MAJOR DISASTER RESPONSE NETWORK MEMBERS

(This list is not inclusive. Feel free to list organizations with whom you interact that are not listed below.)

Community Orgs/Business	**General Emergency**	**Military Organizations (On Base)**
▪ Red Cross ▪ Faith-Based Organizations ▪ Citizen Corps ▪ Salvation Army ▪ Local Schools ▪ Local College/University ▪ Veterinary Services ▪ Professional Associations ▪ Legal Services ▪ Banks ▪ Durable Medical Goods Companies ▪ Other Nonprofit Organizations	▪ EMS ▪ Fire ▪ Police ▪ Community Emergency Response Teams ▪ DMAT ▪ Search and Rescue ▪ State/Local Emergency Response Office ▪ Bioterrorism Agencies ▪ Donations and Volunteer Management Agencies ▪ Hazardous Materials Agencies ▪ Bomb Squad ▪ FEMA (local or regional offices) ▪ Other representatives of state or federal response organizations with whom you do *local* planning and response	▪ Emergency Management ▪ Military Police ▪ Community/Family Support Services ▪ Operations ▪ Media Services ▪ Facilities Management ▪ Installation Safety/Security Forces ▪ Legal Services ▪ Explosive Ordinance Disposal ▪ Public Affairs Office ▪ Office of the Chaplain ▪ Health/Medical ▪ Plans and Programs **Military Organizations (Off Base)** ▪ National Guard ▪ Reserve Units
Infrastructure ▪ Public Works (electric, gas, water) ▪ Public Transit (buses, metro) ▪ Telecommunications Agencies ▪ Transportation Companies ▪ Highway Administration Agencies ▪ Geological Service Agencies ▪ National Weather Service (local affiliate) ▪ Hazardous Material Disposal Agency	**Health** ▪ Hospitals ▪ Mental Health Agencies ▪ Medical Reserve Corps ▪ Mortuary Services ▪ Public Health Dept. ▪ Nursing Homes ▪ Pharmaceuticals ▪ Office of Health Communications ▪ Community Hospitals ▪ State or Local Pharmacy Bureau ▪ VA	**Communications** ▪ Media Organizations (including newspapers, TV, and internet) ▪ Public Affairs Officials (governmental) ▪ Ham Radio Club **Policy** ▪ Legislators ▪ Mayor's Office ▪ Chamber of Commerce ▪ Citizen Action Groups

Thank you for participating in this survey. As mentioned earlier, we will pick it up from you in person during our interview, or we will provide you with a self-addressed envelope to return it to us.

APPENDIX G

Site-Specific Social Network Analysis Findings

In this appendix, we provide descriptions of the emergency management networks at the five sites we visited.

San Antonio Metropolitan Area, Texas, Network

We present the characteristics of the emergency management networks in the San Antonio metropolitan area in two ways: a network diagram (Figure G.1) and table of statistics (Table G.1). In our discussion, we first identify key players in positions of influence or leadership by calculating measures of degree centrality and betweenness centrality for network nodes. We then look at the implications of the network's structure on communications flow, coordination, and innovation. We evaluate the flow of network communications and the potential for coordination (efficiency) and innovation (flexibility) within the network by calculating measures of normalized closeness centralization, density, average path length, and diameter. We then evaluate the network's resiliency, redundancy, and single points of failure, looking again at normed closeness centralization and density, and potential single points of failure as revealed through the betweenness centrality measures.

Figure G.1 presents the key characteristics of the San Antonio metropolitan area network. The green "clouds" represent individual military installations, the VA, and civilian organizations, clustered by city, county or regional, or state organizations. Plain square nodes represent military respondents or organizations—all nodes within the installation clouds are square. Round nodes and hatched square nodes represent civilian respondents and organizations and local representatives of national organizations, respectively. The colors refer to functions. For example, blue nodes represent emergency management functions, and green nodes represent health and medical functions. The size of each node represents its degree centrality.[1] Larger nodes have higher degree centrality scores and, hence, are more influential within the network.[2] Figure G.1 highlights the following key characteristics of the San Antonio metropolitan area network:

---

[1] Degree centrality is measured in terms of the proportion of total possible dyadic connections that each network member has.

[2] Although we tried to select survey respondents who were more likely to play an important role in the larger local disaster preparedness network (from both a military and civilian perspective), and although multiple respondents sometimes nominated other network members, the significance of respondent centrality is partly mitigated by the fact that it was respondents themselves who were responsible for defining their neighborhood networks.

Figure G.1
Combined Military-Civilian Preparedness Network, San Antonio Metropolitan Area, Texas

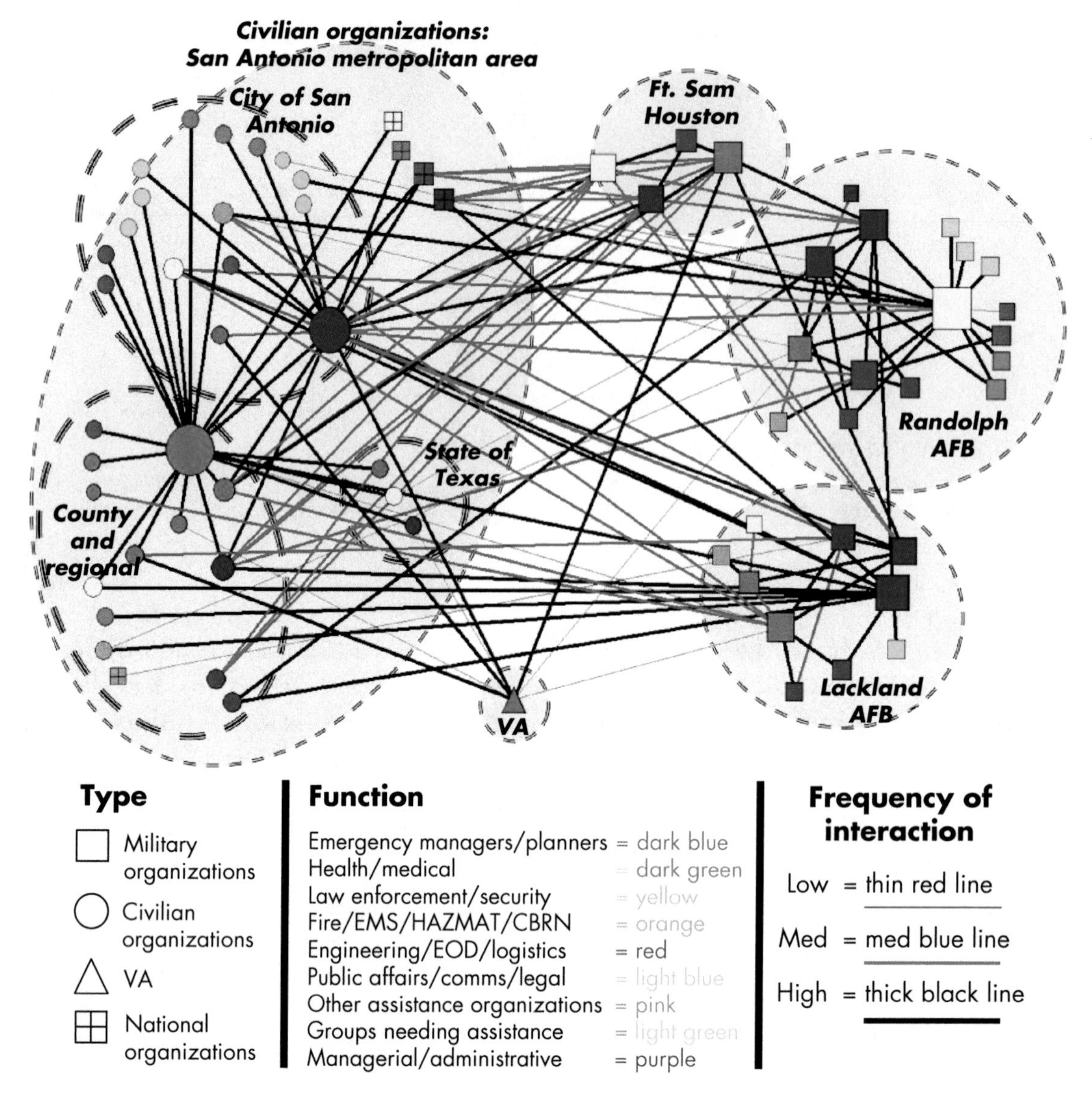

NOTE: EOD = explosive ordnance disposal. CBRN = chemical, biological, radiological, or nuclear.
RAND *TR764-G.1*

- The San Antonio civilian emergency manager (largest dark blue circle) and public health/medical (largest green circle) nodes—located at the center of the "Civilian organizations" cloud—are the largest, indicating that they are more influential than all of the other surrounding nodes.
- The Randolph AFB law enforcement/security node (largest yellow square) is central to the installation's internal network and interacts frequently with counterparts across functions both within the installation and in the civilian community.
- The emergency management/planning (dark blue squares) and medical (green squares) nodes at Randolph AFB, Lackland AFB, and Fort Sam Houston are equally influential within the network and more influential than most of their surrounding nodes.

- At Fort Sam Houston, the emergency management/planning, health/medical, and law enforcement/security (yellow square) nodes are almost equally important to the network.
- Although the most central organizations within each neighborhood tend to connect with like functions across green clouds, their connections are not only limited to similar functions. Ties are still denser within each cloud than across clouds, however.

Key Players

San Antonio Public Health, San Antonio Emergency Management, and Randolph Security Forces are the most influential nodes in the network, reflected in both the size of their nodes in Figure G.1 and their high degree centrality scores, with 64, 54, and 45 percent, respectively, of their total possible relationships in place (Table G.1).

As a reminder, both the network structure illustrated in Figure G.1 and the related measures in Table G.1 are heavily influenced by the methods we used to define the network. The network represents information collected in 16 separate surveys. Organizations nominated by survey respondents rarely had an opportunity to report their own ties. Survey respondents are almost always going to have higher centrality scores than non-respondents. We remain fairly confident, however, that the survey respondents we selected represent the core of the network. As such, we believe that the missing data would trend in the same direction in terms of which organizations proved to be most central to the network.

That public health and the local emergency management office have the largest and potentially most influential roles in the broader emergency preparedness network is in line with our pre-survey hypotheses, as well as with our interview findings. This also correlates with the availability of federal funding. Public health organizations have some of the longest tenures of involvement in emergency preparedness, with state-based morbidity reporting beginning in 1925 as a precursor to today's disease surveillance-programs (Lombardo, 2003). The Federal Civil Defense Act of 1950 (Pub. L. 81-920) formalized the role of state and local governments as having primary responsibility for disaster preparedness and civil defense, although state and local governments had long been informally responsible for both (Arn, 2006, p. 5). The prominent role of Randolph AFB's security forces was also in line with our hypotheses that law enforcement would be among the organizations that formed the core of the local planning and response network.

The size of a network, based on the number of nodes, is another important aspect of its structure because every network and each node in a network has limited resources to use for maintaining network ties and executing assigned tasks (Dunbar, 1992; Hanneman and Riddle,

Table G.1
Network Statistics Summary: San Antonio Metropolitan Area

Node Statistics		Network Statistics			
Highest Degree Centrality	Highest Betweenness Centrality	Network Closeness Centralization (%)	Network Density (%)	Average Path Length	Network Diameter
1. San Antonio PH (64) 2. San Antonio EM (54) 3. Randolph SF (45)	1. San Antonio PH (34) 2. Randolph SF (26) 3. San Antonio EM (24)	33	12	2.84	6

2005, p. 8). The larger the network, the more complex it is likely to be, because of the many different direct and indirect connections that are possible between a large number of nodes.

A large network has both advantages and disadvantages. A large and densely connected network should able to successfully take on more–logistically challenging operations (Enders and Su, 2007, p. 35), but a large network in which connections are haphazard and inconsistent (as might be the case in a network in which many nodes have high betweenness scores) may mean a greater number of opportunities for communications to be lost and for other things to go awry.

In total, survey respondents identified 68 organizations and subunits that comprise the San Antonio metropolitan area emergency preparedness network. Table G.2 breaks down the larger network as determined by the survey respondents. The table first separates network members by location, then by whether it is a civilian or military entity, and finally by function.

Table G.2
Location, Type, and Function of All San Antonio Metropolitan Area Network Participants Identified by Survey Respondents

Network Component	Participants
Organization/unit location	
City of San Antonio, Texas	20
Lackland AFB	10
Randolph AFB	16
Fort Sam Houston	4
Neighboring cities and counties	6
Civilian regional (in-state) or state entities	12
Organization/unit affiliation	
Civilian	37
Military	30
VA	1
Organization/unit function	
Emergency management, plans, exercises	11
Health and medical	12
Security, law enforcement, intelligence	8
Fire, HAZMAT, CBRN	5
Public works, transportation, utilities, engineering	7
Public affairs, communications, media, legal	6
Other support organizations	7
Recipients of assistance	4
Administration, operations, oversight	8

Communications Flow, Coordination, and Innovation

Because the San Antonio metropolitan area network is fairly decentralized, with a normed closeness centralization score of only 33 percent (Table G.1), communications and coordination in this network are likely to be (although not necessarily) less efficient than in a more centralized network.[3] The low density score reinforces the likelihood that communications and coordination are less efficient than they could be. A densely connected network is one in which everyone is closely tied to everyone else. Dense networks tend to communicate regularly and coordinate with one another often, since members both send and receive information via multiple, often overlapping, relationships. The San Antonio metropolitan area's network density score is only 12 percent. This means that only 12 percent of the potential network relationships are actually present—taking into account all network members and not just survey respondents.[4]

Because communications are stovepiped (as can be seen graphically by the number of star-shaped network neighborhoods in which many nodes are connected only to one other central organization), the coordination process is similarly stymied—particularly between network neighborhoods. However, one positive aspect of a less densely connected network structure is the opportunity for innovation. Less dense networks are much more likely to be innovative in coming up with solutions to their problems (Burt, 1992).

Normed closeness centralization is a way to measure how *long* it takes for information to spread throughout the network. The San Antonio metropolitan area network has a normed closeness centralization of about 33 percent. Network members have room to increase their communications speed by increasing the number of ties between potential dyadic pairs. A normed closeness centralization score of 100 percent would mean that everyone in the network could reach everyone else in a single step. But this is not necessarily the most efficient system for regular coordination in a very large network, given the need to maximize one's relationship budget.

Another way to measure closeness, but one that specifies the number of steps between each possible pair of nodes is to measure distance. Average path length measures the average number of steps between all possible pairs of nodes in the network, while the diameter indicates the maximum number of steps between any two network members. The number of steps between nodes is another way to measure how *efficiently* new information or orders are passed through the network, but it can also tell us how efficient network organizations are at maximizing their relationship budgets. Networks with a shorter average path length and diameter should be able to respond more rapidly and efficiently to stimuli with the most efficient use of their time and resources. The average path length in the San Antonio metropolitan area network is 2.84, indicating that the network is fairly efficient in terms of maximizing secondary contacts but could still improve if the target were to reach an average of just two steps between any potential pair of network members. The longest path any node would have to take to reach

---

[3] As a point of clarification, closeness centralization (a network measure) indicates whether there are just a few nodes that seem to control communications with all of the others, whereas the degree (a nodal measure) tells us how much of that control each node has.

[4] Keep in mind that the density score would likely be much higher if information were collected from all the nominated nodes in terms of their own network ties. The centrality of various nodes may also change, but we do not expect the importance of the emergency management, health and medical, or security/law enforcement functions to diminish.

another is six lengths, measured as the diameter of the network. That is not too far, but it is still three times the ideal if our goal is an average of two steps.

Resiliency, Redundancy, and Single Points of Failure

To achieve a better understanding of the potential resiliency of the network, we identify nodes that represent key points of strength and weakness. A highly centralized network is efficient, since the most central node can manage the communications and coordination processes of the network, ensuring that all of its partners receive information and report back as required. However, a highly centralized network is also less resilient, since the most central node is a potential point of critical failure. Network centralization is thus inversely related to network resiliency. High betweenness centrality scores are equated with network "power" based on the desirability of being "between" other pairs of actors in the network. Nodes with high betweenness scores play important broker roles in the network—they are the bridges across which two otherwise unconnected nodes must pass in order to communicate with one another. For example, these powerful nodes can control what each node to which they connect knows about any other node to which they connect. Because the three organizations with the largest ego networks—Public Health, San Antonio Emergency Management, and Randolph's Security Forces—also turn out to be the most "powerful" organizations in the wider San Antonio network based on their measures of betweenness centrality, these are also potential weak points in the network, since the loss of any one of these organizations would severely undermine the network's ability to communicate and coordinate across disconnected units. San Antonio Public Health has the highest betweenness centrality score (34),[5] followed by Randolph's Security Forces Squadron (26) and San Antonio's Emergency Management Office (24).[6]

To better understand the measure of "betweenness," take, for example, the node representing the VA, which lies "between" Lackland AFB's medical (largest green square) node and the Texas State EOC (blue circle inside the "State of Texas" cloud) in Figure G.1. Lackland's medical node is not directly tied to the Texas EOC, but each of them shares a tie with the VA. Pairs for which direct communications are mediated by the presence of a node between them—as is the case with Lackland medical and the Texas EOC—depend on the actor between them to ensure that relevant information is transmitted from one organization to the other, to pass along requests, and so forth. The degree to which pairs of actors have alternative paths (paths that do not go through another node) via which they can access one another decreases their dependency on the node that lies "between" them on the first path. In the case of San Antonio, the VA does not hold a very powerful (central) position in the overall network because it is connected to so few other organizations. But, if the entire network were comprised of only the VA, Lackland medical, and the Texas EOC, then the VA would be the most powerful node in the network.

---

[5] We use the normed value, which is an adjusted score based on the maximum possible betweenness that an actor could have had in this particular network.

[6] Note that, as with all of the cases, we focus primarily on the scores of those organizations that responded to the survey. Because not every organization nominated had the opportunity to define the parameters of its own network, the statistical measures for unrepresented organizations are not as accurate as those for survey respondents.

Norfolk and Virginia Beach Metropolitan Area, Virginia, Network

In the same order as the previous section for the San Antonio metropolitan area, this section presents the characteristics of the emergency management networks in the Norfolk/Virginia Beach area in two ways: a picture (Figure G.2) and a table of statistics (Table G.3). In our discussion, we again identify key players in positions of influence or leadership by calculating measures of degree centrality and betweenness centrality for network nodes. We then look at the implications of the network's structure on communications flow, coordination, and innovation. We evaluate the flow of network communications and the potential for coordination (efficiency) and innovation (flexibility) within the network by calculating measures of closeness centrality, density, and reach efficiency. We then evaluate the network's resiliency, redundancy, and single points of failure, looking again at closeness centralization and density, and then examining potential single points of failure as revealed through the betweenness centrality measures.

Also similar to the previous case, the network structure (presence of so many star-shaped neighborhoods) illustrated in Figure G.2 and the related measures discussed below are heavily influenced by the method used to define the network. Nodes are again sized in accordance with a measure of degree centrality; larger nodes indicate organizations that are more central to the network and that wield the most influence. And neighborhoods are labeled in accordance with their broad affiliation (city, VA, specific military installation).

Figure G.2 presents the key characteristics of the Norfolk/Virginia Beach metropolitan area network. As a reminder, the green "clouds" represent individual military installations, the VA, and civilian organizations, clustered by city, county or regional, or state organizations. Plain square nodes represent military respondents or organizations, while round nodes and hatched square nodes represent civilian respondents or organizations and local representatives of national organizations, respectively. A key to the different colors (representing functions) is provided beneath Figure G.2. The size of the nodes represents their degree centrality.[7] Larger nodes have higher degree centrality scores and, hence, are more influential within the network.[8] Figure G.2 highlights the following key characteristics of the Norfolk/Virginia Beach metropolitan area network:

- The Norfolk and Virginia Beach civilian emergency managers (largest dark blue circles), Virginia Beach Public Health (largest green circle), and the Virginia Beach Fire Department (largest orange circle) are the largest civilian nodes, indicating that they are the most influential civilian nodes.
- NS Norfolk emergency management office (largest dark blue square) and NMC Portsmouth (largest dark green circle) are similarly influential to the Virginia Beach civilian emergency management office (the larger of the two blue circles).
- NS Norfolk Fire Department (large orange square) is similarly influential to the Virginia Beach Fire Department (largest orange circle) and to the Norfolk civilian emer-

[7] Degree centrality is measured in terms of the proportion of total possible dyadic connections that each network member has.

[8] Although we tried to select survey respondents who were more likely to play an important role in the larger local disaster preparedness network (from both a military and civilian perspective), and although multiple respondents sometimes nominated other network members, the significance of respondent centrality is partly mitigated by the fact that it was respondents themselves who were responsible for defining their neighborhood networks.

Figure G.2
Combined Military-Civilian Preparedness Network, Norfolk/Virginia Beach Metropolitan Area, Virginia

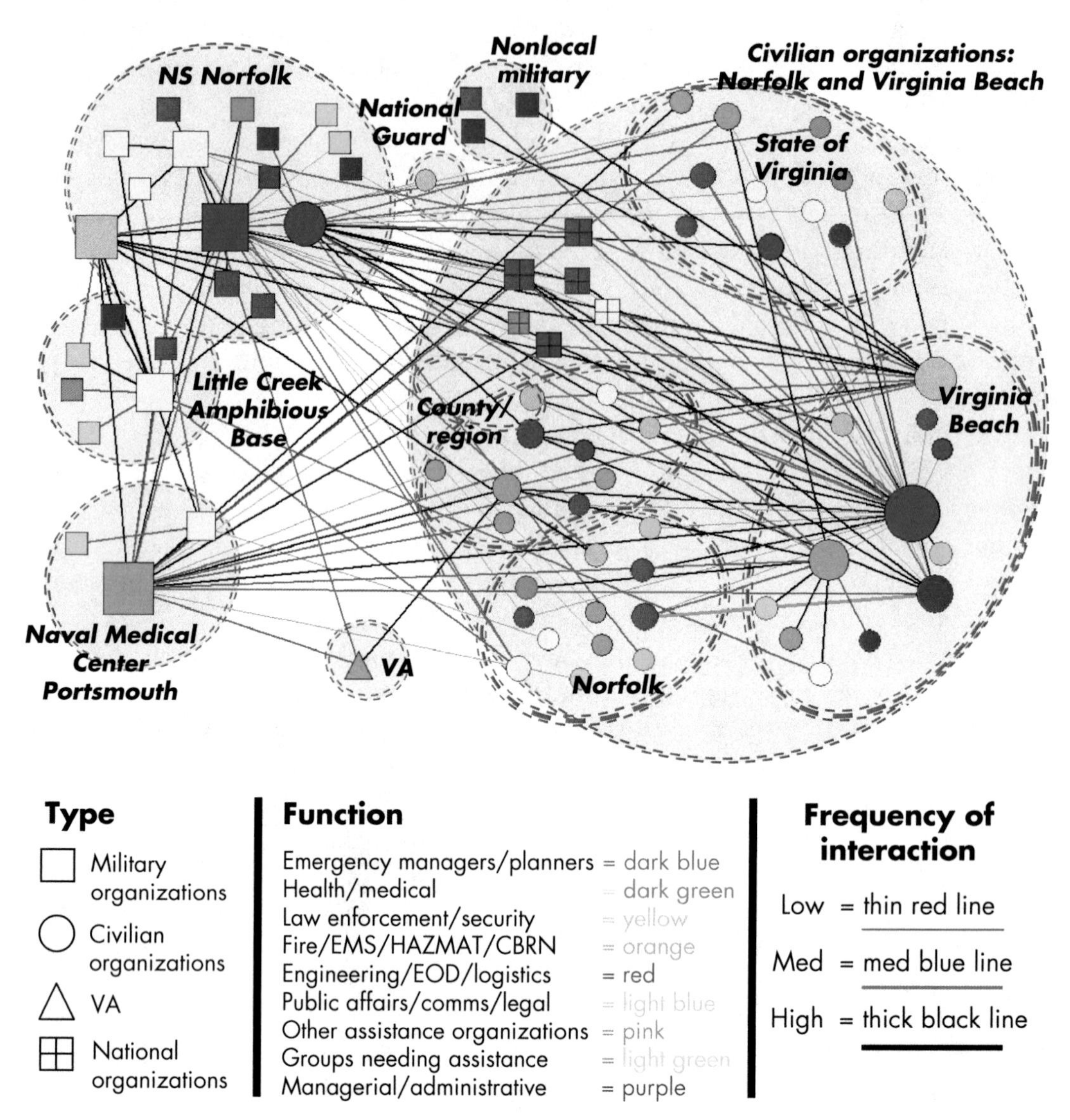

RAND *TR764-G.2*

gency management office (smaller of the two large blue circles) and Virginia Beach Public Health (largest green circle).

- Each of the most influential nodes is connected to at least one, but usually more, of the other most influential nodes, which helps to bridge the divide across both civilian and military and differing functional communities.
- The densest inter-organizational ties are in the public health/medical community (green squares, circles, and triangle). The star-shaped neighborhoods are connected to each other via multiple ties.

Key Players

As Figure G.2 and Table G.3 illustrate, the Virginia Beach Emergency Management Office, NMC Portsmouth, and NS Norfolk's EOC have the highest degree centrality scores, with 56, 48, and 41 percent, respectively, of their possible dyadic connections in place. As a reminder, the higher the degree score, the more likely that node is to be influential within the network. This is again not surprising, as we hypothesized that emergency management and health and medical offices would be the most influential in the broader local network. The City of Norfolk's Emergency Management Office is next in line, with a degree score of 36.

The Norfolk/Virginia Beach network displayed in Figure G.2 represents the information collected in 14 separate surveys. In total, survey respondents identified some 81 organizations and subunits that comprised the Norfolk/Virginia Beach network. Table G.4 breaks down the larger network as determined by survey respondents. The table first separates network members by location, then by whether it is a civilian or military entity, and finally by function.

Communications Flow, Coordination, and Innovation

As shown in Tables G.1 and G.3, the Norfolk/Virginia Beach network is slightly more centralized and slightly less dense than the San Antonio metro area network, with a closeness centralization score of 39 percent[9] (versus San Antonio's 33 percent) and a density score of 9 percent (versus San Antonio's 12 percent). Interestingly, based on the information we collected, the Norfolk/Virginia Beach network is almost 20 percent larger than the San Antonio metropolitan area network—the former consists of 81 nodes, compared to San Antonio's 68 nodes (see Table 5.3 in Chapter Five)—and this is probably an understatement, given the complexity of the Norfolk/Virginia Beach area. Given the tendency for an inverse relationship between density and network size, we might have expected to see a greater difference between the two networks. As with San Antonio, communications and coordination in this network are likely to be less efficient than in a more densely connected network in which organizations are more closely tied to everyone else in the network. The network density score of only 9 percent means that only about 9 percent of the potential network relationships are actually present in

Table G.3
Network Statistics Summary: Norfolk and Virginia Beach Metropolitan Area, Virginia

Node Statistics		Network Statistics			
Highest Degree Centrality	Highest Betweenness Centrality	Network Closeness Centralization (%)	Network Density (%)	Average Path Length	Network Diameter
1. Virginia Beach EM (56) 2. NMC Portsmouth (48) 3. NS Norfolk EOC (41)	1. NMC Portsmouth (33) 2. Virginia Beach EM (32) 3. NS Norfolk EOC (30)	39	9	2.81	5

[9] Indicating that about two-fifths of the network members can reach everyone else within a single step.

Table G.4
Location, Type, and Function of All Norfolk/Virginia Beach Metropolitan Area Network Participants Identified by Survey Respondents

Network Component	Participants
Organization/unit location	
Norfolk, Virginia	10
Virginia Beach, Virginia	6
Hampton Roads/Tidewater region	18
Other cities or counties	2
State entities	11
NS Norfolk	16
NMC Portsmouth	3
NAB Little Creek	8
Nonlocal military organization	2
Organization/unit affiliation	
Civilian	47
Military	29
VA	1
Organization/unit function	
Emergency management, plans, exercises	12
Health and medical	12
Security, law enforcement, intelligence	13
Fire, HAZMAT, CBRN	10
Public works, transportation, utilities, engineering	13
Public affairs, communications, media, legal	6
Other support organizations	5
Recipients of assistance	2
Administration, operations, oversight	4

Norfolk/Virginia Beach. This measure takes into account all reported network members and not just survey respondents.[10]

Based on the network structure alone, one would conclude that it is difficult for network members to coordinate efficiently within and across the network. It should be easier for those involved in the slightly more densely connected health/medical function (see Figure G.2). The

[10] The density score will likely be much higher as we obtain more information about the nominated nodes and their own network ties. The centrality of various nodes may also change, but we do not expect the importance of the emergency management or health and medical functions to diminish significantly.

one positive aspect of this dispersed network structure is the opportunity for innovation, but the lack of efficient communications tends to force organizations to rely on ad hoc solutions because they lack awareness of existing, tested solutions for emergency situations.

In terms of offsetting the inefficiencies of a more loosely connected network, the Norfolk network is similarly efficient to the San Antonio network in terms of maximizing secondary contacts, with an average distance of 2.81. That means that the Norfolk network is similarly efficient to the San Antonio network, which has an average path length of 2.84, in terms of maximizing secondary contacts. That means that the Norfolk network is fairly closely connected despite its large size. The longest path between any two nodes is five lengths, measured as the diameter of the network. But network members still have room to increase their communications speed by more than half if they increase the number of ties between potential dyadic pairs, or they could at least maximize their efficiency by increasing the number of ties between potential pairs until the average distance of the network equals two.

Resiliency, Redundancy, and Single Points of Failure

The centralization in the Norfolk/Virginia Beach network renders it less resilient to disruption, since members of each of the star-shaped neighborhoods would be completely cut off from the broader local preparedness network if something were to happen to the central node in any of their neighborhoods. If more complete data collection were to indicate greater decentralization than centralization in the overall network, there may be some hidden resiliency that we cannot currently measure. This assumes that responsibilities must be spread out across the broader network, since there is not enough centralization to account for narrower control of responsibilities.

The same three organizations with the largest degree centrality scores—Virginia Beach Emergency Management, NS Norfolk's EOC, and NMC Portsmouth—also turn out to be the most powerful organizations in the wider Norfolk network based on the high values of their betweenness centrality measures. These organizations are also therefore potential weak points in the network, since the loss of any one of these organizations would severely undermine the network's ability to communicate and coordinate across disconnected units. NMC Portsmouth has a slightly higher betweenness centrality score (33 percent),[11] followed by Virginia Beach Emergency Management (32 percent), and Norfolk EOC (30 percent). As a reminder, organizations with high betweenness scores are the bridges across which otherwise unconnected pairs can reach one another.

City of Columbus and Muscogee County, Georgia, Network

This section presents the characteristics of the emergency management networks in the City of Columbus and Muscogee County through a network diagram (Figure G.3) and a table of statistics (Table G.5). We again identify key players in positions of influence or leadership by calculating measures of degree centrality and betweenness centrality for network nodes. We then look at the implications of the network's structure on communications flow, coordination, and innovation. We evaluate the flow of network communications and the potential for

[11] We use the normed value, which is an adjusted score based on the maximum possible betweenness that an actor could have had in this particular network.

Figure G.3
Combined Military-Civilian Preparedness Network, City of Columbus and Muscogee County, Georgia

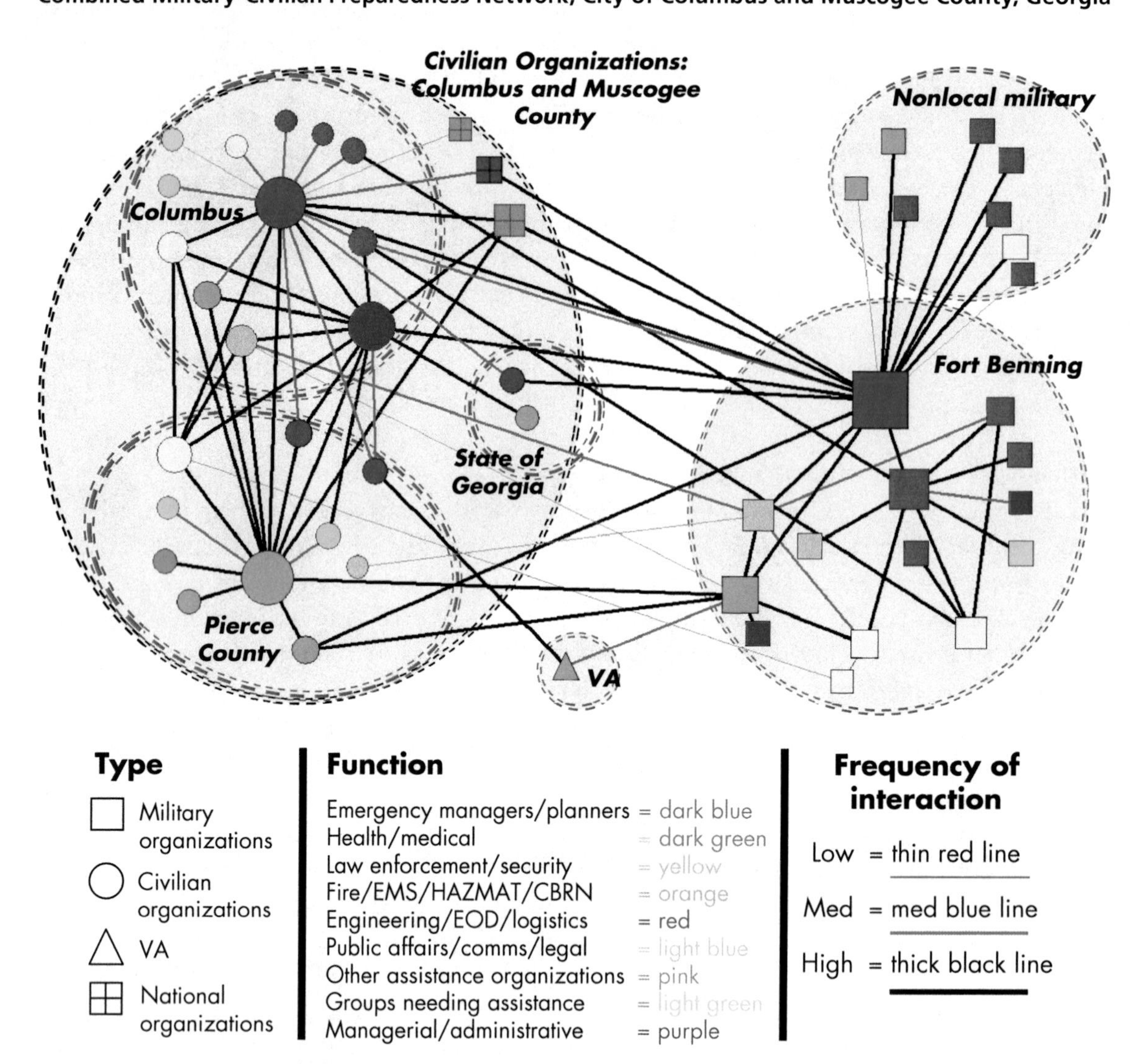

RAND *TR764-G.3*

Table G.5
Network Statistics Summary: City of Columbus/Muscogee County, Georgia

Node Statistics		Network Statistics			
Highest Degree Centrality	Highest Betweenness Centrality	Network Closeness Centralization (%)	Network Density (%)	Average Path Length	Network Diameter
1. Fort Benning EM (49) 2. Muscogee County Public Health (45) 3. Columbus EM (42)	1. Fort Benning EM (44) 2. Columbus EM (37) 3. Muscogee County Public Health (21)	39	11	2.77	5

coordination (efficiency) and innovation (flexibility) within the network by calculating measures of closeness centralization, density, and reach efficiency. We then evaluate the network's resiliency, redundancy, and single points of failure, looking again at closeness centrality and density, and then examining potential single points of failure as revealed through our betweenness centrality measures.

As with both prior cases (San Antonio and Norfolk/Virginia Beach), the network structure and related measures are heavily influenced by the method used to define the network. Nodes are again sized in accordance with a measure of degree centrality; larger nodes indicate organizations that are more central to the network and that wield the most influence. And neighborhoods are labeled in accordance with their broad affiliation (city, VA, military installation). A key to the different node colors (representing functions) and shapes (representing organization) is provided beneath Figure G.3. The figure highlights the following key characteristics of the City of Columbus/Muscogee County network:

- The Fort Benning emergency management office (largest dark blue square), the Muscogee County Department of Public Health (largest green circle), Columbus's emergency management office, and the Office of Homeland Security at City Hall (large purple circle) are the most influential nodes in the network.
- The Fort Benning emergency management office (largest dark blue square) is also the military organization with the largest number of ties to civilian organizations.
- The VA (green triangle) provides a bridging tie between Martin Army Community Hospital (largest green square) and emergency management offices from other Georgia counties (small blue circle directly connected to the VA).
- MMRS (small green circle at the bottom of the civilian cloud) provides redundant ties between public health (largest green circle) and Fort Benning emergency management and between public health and Martin Army Community Hospital—establishing some resiliency across the broader network.

Key Players

As Figure G.3 and Table G.5 illustrate, nodes with the highest degree centrality measures were the Fort Benning Emergency Management Office, the Muscogee County Public Health Department, and Columbus Emergency Management Office, with 49, 45, and 42 percent, respectively, of their possible dyadic connections in place. This means that these three organizations are the most influential within the broader Columbus/Muscogee County network. This is again, not surprising, as we hypothesized that emergency management and health and medical offices would be the most influential in the broader network. The Columbus Office of Homeland Security is next in line with a degree centrality measure of 38.

The City of Columbus/Muscogee County network displayed in Figure G.3 represents the information collected in 13 separate surveys. In total, survey respondents identified 54 organizations and subunits that comprised the City of Columbus/Muscogee County network. Table G.6 breaks down the larger network as determined by survey respondents. The table first separates network members by location, then by whether it is a civilian or military entity, and finally by function.

Table G.6
Location, Type, and Function of All City of Columbus/Muscogee County, Georgia, Network Participants Identified by Survey Respondents

Network Component	Participants
Organization/unit location	
Columbus, Georgia, and Muscogee County	25
Fort Benning	14
Neighboring cities and counties	3
State of Georgia	3
State of Alabama	1
Nonlocal military units	8
Organization/unit affiliation	
Civilian	31
Military	22
VA	1
Organization/unit function	
Emergency management, plans, exercises	6
Health and medical	11
Security, law enforcement, intelligence	8
Fire, HAZMAT, CBRN	4
Public works, transportation, utilities, engineering	13
Public affairs, communications, media, legal	2
Other support organizations	3
Recipients of assistance	4
Administration, operations, oversight	3

Communications Flow, Coordination, and Innovation

The Columbus/Muscogee County network is similarly structured to that in the Norfolk/Virginia Beach area in terms of centralization (with a closeness centralization score of 39 percent in Columbus/Muscogee County, as shown in Table G.5, the same as in Norfolk/Virginia Beach). The Columbus/Muscogee County network is the smallest of the three cases so far, comprised of 54 organizations versus Norfolk/Virginia Beach's 81 and San Antonio's 68. As with San Antonio and Norfolk/Virginia Beach, communications and coordination in this network are likely to be less efficient than they would be in a more densely connected network. The City of Columbus/Muscogee County's network-density score is 11 percent, so only 11 percent of the potential network relationships are actually present. Communications and coordination across the Columbus/Muscogee County network are approximately the same as in the San Antonio network, and both have significant room for improvement.

The level of efficiency of the Columbus/Muscogee County network is similar to those in San Antonio and Norfolk/Virginia Beach in terms of maximizing secondary contacts, with an average path length of 2.77, compared with 2.84 in San Antonio and 2.81 in Norfolk/Virginia Beach. Similar to Norfolk/Virginia Beach, the diameter is 5. Average path length measures the percent of nodes that are within two steps of each other. Columbus/Muscogee County survey respondents score quite high in this area. Those nodes with lower scores have room to improve their efficiency in reaching everyone else in the network in fewer steps than is presently the case. Although the longest path between any two nodes is five lengths, measured as the diameter of the network, the Columbus/Muscogee County network is still a fairly closely connected one, based both on the average path length and on closeness. As in Norfolk/Virginia Beach, Columbus/Muscogee County's network members could increase their communications speed by more than half if they increase the number of ties between potential dyadic pairs, or they could aim to maximize efficiency by increasing the number of ties between potential pairs until the average path length of the network equals two.

Resiliency, Redundancy, and Single Points of Failure

As noted previously, Figure G.3 indicates that there are a few organizations with redundant ties in the Columbus/Muscogee County network, such as Fort Benning's Emergency Management Office and Muscogee County Public Health, which have redundant ties in the health/medical community due to their MMRS participation. Because we are equating redundancy with resiliency, the organizations with the largest number of redundant ties are the ones that might provide continuity in the larger network in case of a disaster. Other than these few nodes, however, the local Columbus/Muscogee County network is no more redundant overall than those in San Antonio or Norfolk and will not therefore be very resilient to the removal or incapacitation of key network members.[12]

As in the two preceding cases, the same three organizations with the largest degree centrality scores are also the most powerful organizations in the wider Columbus/Muscogee County network based on their measures of betweenness centrality: Fort Benning Emergency Management Office, Columbus Emergency Management Office, and the Muscogee County Public Health Department have betweenness scores of 44, 37, and 21, respectively (Table G.5). These organizations are also therefore potential weak points in the network, since the loss of any one of these organizations would severely undermine the network's ability to communicate and coordinate across disconnected units. This is because they are currently the only connectors between multiple pairs who would not otherwise have any contact at all.

City of Tacoma and Pierce County, Washington, Network

This section presents the characteristics of the emergency management networks in the City of Tacoma/Pierce County, Washington, network illustrated by a picture (Figure G.4) and a table of statistics (Table G.7). We identify key players in positions of influence or leadership by calculating measures of degree centrality and betweenness centrality for network nodes. We evaluate the flow of network communications and the potential for coordination (efficiency) and innovation (flexibility) within the network by calculating measures of closeness centrality,

[12] We lack sufficient data to use more direct measures of redundancy.

Figure G.4
Combined Military-Civilian Preparedness Network, City of Tacoma and Pierce County, Washington

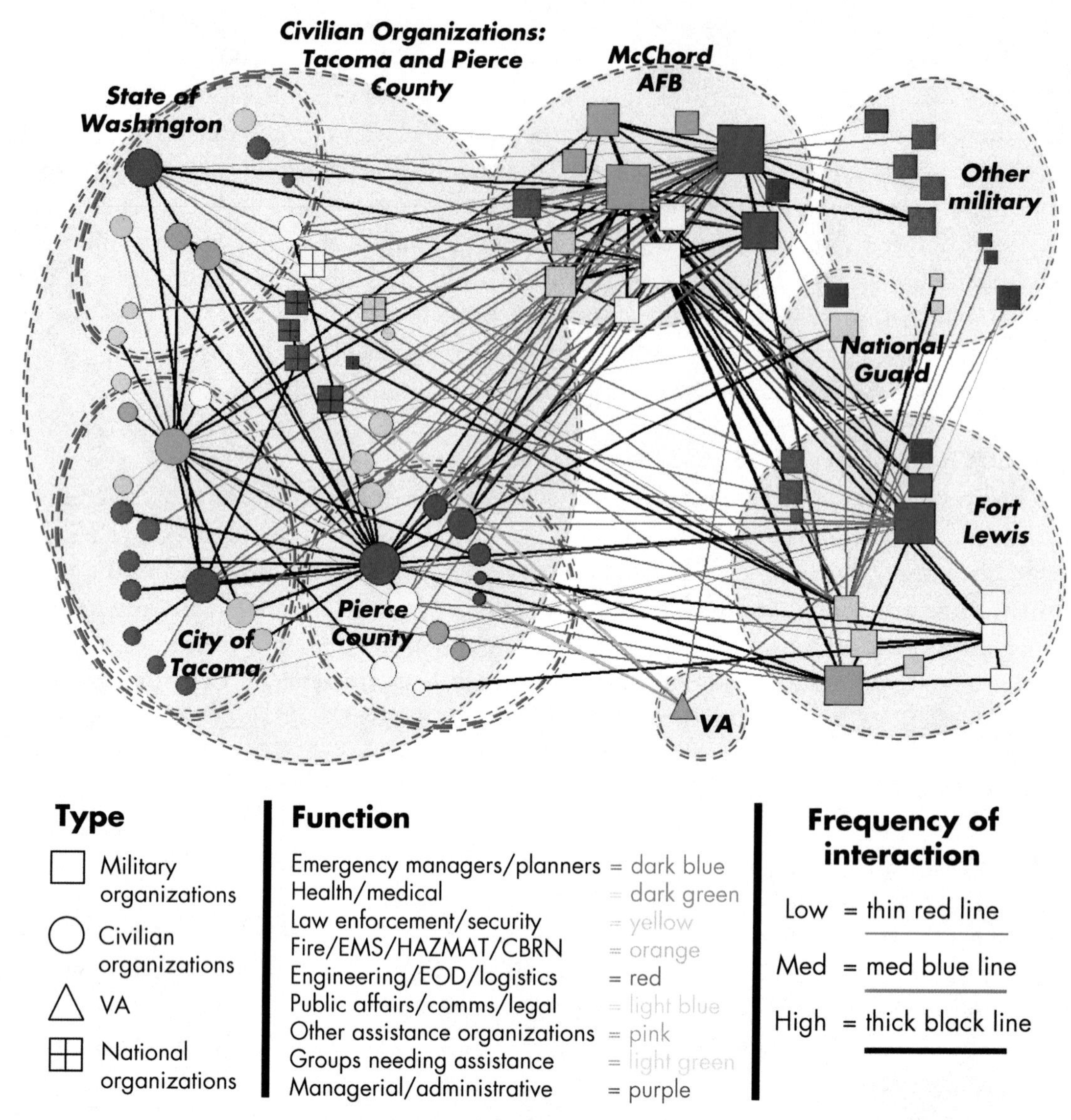

RAND *TR764-G.4*

density, and reach efficiency. We then evaluate the network's resiliency, redundancy, and single points of failure, looking again at closeness centrality and density, and then examining potential single points of failure as revealed through our betweenness centrality measures.[13]

Nodes are again sized in accordance with a measure of degree centrality; larger nodes indicate organizations that are more central to the network and that wield the most influence. And neighborhoods are labeled in accordance with their broad affiliation (city, VA, military installation). A key to the different node colors (representing functions) and shapes (represent-

[13] Remember that the network structure and related measures are heavily influenced by the method used to define the network—leading to a greater than usual number of star-shaped networks.

Table G.7
Network Statistics Summary: City of Tacoma/Pierce County, Washington

Node Statistics		Network Statistics			
Highest Degree Centrality	**Highest Betweenness Centrality**	**Network Closeness Centralization (%)**	**Network Density (%)**	**Average Path Length**	**Network Diameter**
1. McChord Plans (23) 2. McChord Medical (20) 3. Pierce County EM (19)	1. McChord Plans (39) 2. Fort Lewis Installation Safety Office (18) 3. Pierce County OEM (17)	41	9	2.75	5

ing organization) is provided beneath Figure G.4. The figure highlights the following key characteristics of the City of Tacoma/Pierce County network:

- The most influential nodes in the network are the Pierce County emergency management office (largest dark blue circle), the McChord Planning Office (largest dark blue square), and McChord's medical flight (largest green square).
- Tacoma/Pierce County Public Health (largest green circle), Madigan Army Medical Center (largest green in the Fort Lewis cloud), and McChord's antiterrorism office are also very influential.
- The Tacoma/Pierce County network seems to have the densest pattern of ties across neighborhoods that we have seen in any of our cases. This density is apparent based on the figure but is not well reflected in the statistics (shown in Table G.7) because of the skewing effect of so many pendant nodes.
- Survey respondents nominated more nonlocal military organizations as being important to their emergency preparedness networks than in any other case. These are the blue and red squares located in the "Other military" cloud and correspond to USNORTHCOM (blue square), in addition to U.S. Army North, U.S. Air Force Civil Engineer Support Agency, U.S. Air Force Mobility Command, and the Air Staff's Readiness and Emergency Management Office. We believe these nominations to be largely a function of McChord's active role in ongoing overseas conflicts.

Key Players

As Figure G.4 and Table G.7 illustrate, nodes with the highest degree centrality measures were the McChord Plans and Programs Office, McChord Medical, and the Pierce County Office of Emergency Management, with 23, 20, and 19 percent, respectively, of their possible dyadic connections in place. This means that these three organizations are the most influential within the broader Tacoma/Pierce County network—although these degree centrality measures are far less impressive than in the three preceding cases—at about half the level of influence of the top three nodes in the San Antonio, Norfolk/Virginia Beach, and Columbus/Muskogee County networks (which ranged from 64 to 49, as shown in Tables G.1, G.3, and G.5).

The Tacoma/Pierce County network displayed in Figure G.4 represents the information collected in 16 separate surveys. In total, survey respondents identified 86 organizations and subunits that comprise the Tacoma/Pierce County network. Table G.8 breaks down the larger

Table G.8
Location, Type, and Function of All City of Tacoma and Pierce County, Washington, Network Participants Identified by Survey Respondents

Network Component	Participants
Organization/unit location	
Tacoma and Pierce County, Washington	32
Seattle and King County, Washington	2
Nearby cities and counties, Washington	8
State of Washington	8
McChord AFB	15
Fort Lewis	14
Nonlocal military organizations	8
Organization/unit affiliation	
Civilian	47
Military	38
VA	1
Organization/unit function	
Emergency management, plans, exercises	14
Health and medical	11
Security, law enforcement, intelligence	12
Fire, HAZMAT, CBRN	15
Public works, transportation, utilities, engineering	20
Public affairs, communications, media, legal	1
Other support organizations	1
Recipients of assistance	3
Administration, operations, oversight	8

network as determined by survey respondents. The table first separates network members by location, then by whether it is a civilian or military entity, and finally by function.

Communications Flow, Coordination, and Innovation

The Tacoma/Pierce County network is actually the largest of the four cases so far, comprised of 86 organizations, although, by population size, it is one of the smallest cases. The Tacoma/Pierce County network is similar in density to the Norfolk network, with only about 9 percent of the potential network relationships present. As with all of the preceding cases, there is room for improved efficiency in the areas of communications and coordination.

The longest distance between any two nodes is five lengths, presented as the diameter of the network, with an average of just 2.75 steps between nodes. Tacoma/Pierce County's closeness measure (41 percent) (Table G.7) is not far from Norfolk/Virginia Beach's (39 percent)

(Table G.3), but it is still the highest across the five sites we studied. This means that the speed of communications in the Tacoma/Pierce County network is potentially the fastest of any of the other networks, although members could still increase that speed by increasing the number of ties between potential dyadic pairs or increase efficiency by lowering the average distance to two.

Resiliency, Redundancy, and Single Points of Failure

Although not well reflected in Table G.7, the Tacoma/Pierce County network seems to have the densest pattern of ties across neighborhoods that we have seen in any of the five networks we studied. It is the skewing effect of so many pendant nodes that prevents us from seeing this density reflected in the table. This density is more apparent if we look at Figure G.4. The figure shows a large number of redundant ties between these organizations. The number of redundant ties suggests that Tacoma/Pierce County might be one of the most resilient networks. To further increase the resiliency of the network, it would need to establish even more redundant ties among organizations. The potentially weak points in the Tacoma/Pierce County network, based on their betweenness scores, are McChord Plans (39), Fort Lewis Installation Safety Office (18), and Pierce County's Office of Emergency Management (17).

City of Las Vegas and Clark County, Nevada, Network

This section presents the characteristics of the emergency management networks in the City of Las Vegas/Clark County network illustrated by a picture (Figure G.5) and a table of statistics (Table G.9). We identify key players in positions of influence or leadership by calculating measures of degree centrality and betweenness centrality for network nodes. We evaluate the flow of network communications and the potential for coordination (efficiency) and innovation (flexibility) within the network by calculating measures of closeness centrality, density, and reach efficiency. We then evaluate the network's resiliency, redundancy, and single points of failure, looking again at closeness centrality and density, and then examining potential single points of failure as revealed through our betweenness centrality measures.[14]

Nodes are again sized in accordance with a measure of degree centrality; larger nodes indicate organizations that are more central to the network and that wield the most influence. And neighborhoods are labeled in accordance with their broad affiliation (city, VA, military installation). A key to the different node colors (representing functions) and shapes (representing organization) is provided beneath Figure G.5. The figure highlights the following key characteristics of the City of Las Vegas/Clark County network:

- There are a number of similarly sized—and therefore influential—nodes in the network, including the Bioenvironmental Engineering Flight, in the 99th Aerospace Medicine Squadron Nellis AFB (largest dark green square), the MOFH at Nellis AFB (second largest dark green square), Southern Nevada Public Health (largest dark green circle), the Las Vegas Police Department (largest yellow circle), the Clark County Office of Emergency

[14] A reminder that the network structure and related measures are heavily influenced by the method used to define the network—leading to more than usual number of star-shaped networks.

Figure G.5
Combined Military-Civilian Preparedness Network, City of Las Vegas and Clark County, Nevada

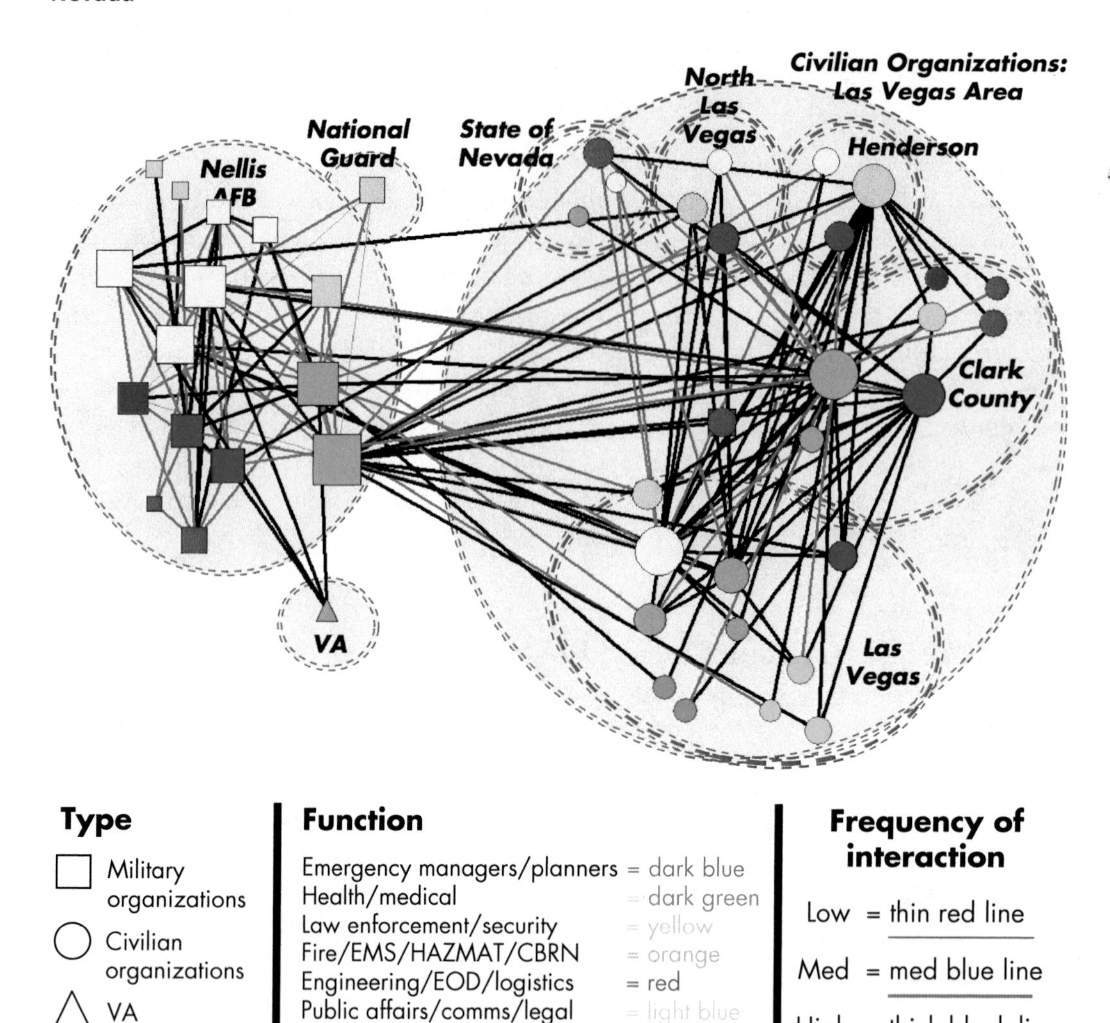

Type	Function		Frequency of interaction
Military organizations	Emergency managers/planners	= dark blue	Low = thin red line
Civilian organizations	Health/medical	= dark green	Med = med blue line
VA	Law enforcement/security	= yellow	High = thick black line
National organizations	Fire/EMS/HAZMAT/CBRN	= orange	
	Engineering/EOD/logistics	= red	
	Public affairs/comms/legal	= light blue	
	Other assistance organizations	= pink	
	Groups needing assistance	= light green	
	Managerial/administrative	= purple	

RAND *TR764-G.5*

Management (largest dark blue square), and the Henderson Fire Department (largest orange circle).

- Based mainly on inspection of the network diagram and borne out in Tables G.9 and 5.3 (the latter in Chapter Five), the Nellis AFB network is an extremely dense network—perhaps the densest neighborhood we have seen.
- The VA is connected to the Nellis AFB network but is completely disconnected from the civilian community.

Key Players

As Figure G.5 and Table G.9 illustrate, nodes with the highest degree centrality scores were the Southern Nevada Health District, the Las Vegas Emergency Preparedness Coordinator, and Nellis AFB's BEE with 61, 56, and 52 percent, respectively, of their possible dyadic connections in place. These are very high degree scores and similar to the degree scores of the most influential members in the San Antonio network. They are almost three times as high as the degree scores in the Tacoma network. High degree scores mean that these three organizations are the most influential within the broader City of Las Vegas/Clark County network.

Figure G.5 indicates that there is much more redundancy in the City of Las Vegas/Clark County network than in the other networks, which should indicate greater overall resiliency for the network. The most remarkable thing about the Las Vegas network is that only one organization was identified by a survey respondent without any duplication on the part of other survey respondents. The City of Las Vegas/Clark County network represents a far more complete network than any of the preceding cases described in this chapter. This may be an artifact of survey design (and completeness of reporting from this and other sites), or it may reflect the true nature of that network. Because of this fact, we have more confidence in the accuracy of both the network and the node measures for this case.

The City of Las Vegas/Clark County network displayed in Figure G.5 represents the information collected in 11 separate surveys. In total, survey respondents identified 45 organizations and subunits that comprise the City of Las Vegas/Clark County network. Table G.10 breaks down the larger network as determined by survey respondents. The table first separates network members by location, then by whether they are a civilian or military entity, and finally by function.

Communications Flow, Coordination, and Innovation

The Las Vegas/Clark County network is the densest of our five cases, with a density measure (16) that is almost twice those of the Norfolk/Virginia Beach and Tacoma/Pierce County networks (both about 9 percent). The Las Vegas/Clark County network is also the smallest of our four cases, comprised of 45 organizations, just more than half the size of Tacoma, the largest network. Although there is still room for improved efficiency of communications and coordination, the Las Vegas/Clark County network should be the most efficient of our five networks in terms of communications and coordination. Table G.9 summarizes statistical measures for the Las Vegas/Clark County network.

The Las Vegas/Clark County network is the most efficient network in terms of maximizing secondary contacts, with an average distance of 2.12. The longest path between any two nodes is just four lengths, the diameter of the network—and the lowest measure of any of the networks. Las Vegas/Clark County's closeness measure is 36 percent (higher than San Anto-

Table G.9
Network Statistics Summary: City of Las Vegas and Clark County, Nevada

Node Statistics		Network Statistics			
Highest Degree Centrality	Highest Betweenness Centrality	Network Closeness Centralization (%)	Network Density (%)	Average Path Length	Network Diameter
1. Southern Nevada Health District (61) 2. Las Vegas Emergency Preparedness (56) 3. Nellis BEE (52)	1. Southern Nevada Health District (21) 2. Las Vegas Emergency Prep (17) 3. Nellis BEE (16)	36	16	2.12	4

Table G.10
Location, Type, and Function of All City of Las Vegas and Clark County, Nevada, Network Participants Identified by Survey Respondents

Network Component	Participants
Organization/unit location	
Las Vegas, Nevada	15
Clark County, Nevada	6
North Las Vegas, Nevada	3
Henderson, Nevada	3
State of Nevada	2
Nellis AFB	16
Organization/unit affiliation	
Civilian	28
Military	16
VA	1
Organization/unit function	
Emergency management, plans, exercises	4
Health and medical	9
Security, law enforcement, intelligence	9
Fire, HAZMAT, CBRN	6
Public works, transportation, utilities, engineering	6
Public affairs, communications, media, legal	6
Other support organizations	2
Recipients of assistance	3
Administration, operations, oversight	3

nio but lower than Columbus/Muscogee County), so the speed of communications in the Las Vegas network is moderately rapid. As with all of the other networks, however, members could still increase that speed by increasing the number of ties between potential dyadic pairs.

Resiliency, Redundancy, and Single Points of Failure

The level of redundant ties in the Las Vegas/Clark County network is very high overall (Figure G.5). The potentially weak points in the Las Vegas/Clark County network, based on their betweenness scores, are the same organizations that wield the most influence in the network—the Southern Nevada Health District (21), the Las Vegas Emergency Preparedness Office (17), and Nellis's Bioenvironmental Engineering Unit (16) (Table G.9).

APPENDIX H

Other Preparedness Support Tools and Methods

This appendix contains tools and methods other than ours, for risk assessment, planning, event management, and multiple phases (Tables H.1–H.4, respectively).

Table H.1
Risk Assessment Tools and Methods

Name (Owner)	Access	Hazards	Scope	User Input	Output	Approach	Target Clientele
Kaiser Hazard Vulnerability Analysis (Kaiser Permanente)	Public	All	Risk assessment by category of hazard	Knowledge about potential risks to infrastructure or population under consideration	Calculated risk for spectrum of hazards	Worksheet filled out by user assigning probability and severity scores to hazards in order to calculate a risk score	Hospitals
Source: http://www.osha.gov/dts/osta/bestpractices/html/hospital_firstreceivers.html							
Joint Staff Integrated Vulnerability Assessments (JSIVA) (conducted by Defense Threat Reduction Agency [DTRA] teams)	Military only	All	Installation-specific assessment of force and critical infrastructure protection	DTRA teams	Gaps in protection of installation forces and critical infrastructure	Conducted by 6 DTRA teams of 7 people each	Military installations, all services
Source: Web page no longer available online.							
Hazards U.S. Multi-Hazard (HAZMUS-MH) (Federal Emergency Management Agency [FEMA])	Public	Natural: earthquakes, hurricane winds, floods	Assesses potential losses from floods, hurricane winds, and earthquakes	Demographics, building stock, essential facilities (hospitals, police, fire, schools), transportation and utility lifelines, casualties, shelter, building, and others	Estimation of potential future losses: maps and tables	Estimates physical and economic consequences of natural hazards	Best for regional analysis (local governments) but can also be applied to single buildings or infrastructure
Source: http://www.fema.gov/plan/prevent/hazus/							

Table H.1—Continued

Name (Owner)	Access	Hazards	Scope	User Input	Output	Approach	Target Clientele
Electronic Mass Casualty Assessment and Planning Scenarios (EMCAPS) (Johns Hopkins Office of Critical Event Preparedness and Response)	Public	Intentional: chemical, biological, radiological, nuclear, or explosive (CBRNE)	Casualty estimates for various scenarios	Estimates about local population density, wind speed, and other local conditions	Casualty estimates	Modeling of disaster scenarios for drill planning. In the model, select one of the scenarios and adjust inputs (e.g., bomb size, population density, quantity of release, wind speed)	Local governments
Source: http://www.hopkins-cepar.org/EMCAPS/EMCAPS.html							
Consequence Assessment Tool Set (CATS) (SAIC)	Federal, state, and local emergency responders	All	Predicts damages and assesses consequences of manmade and natural disasters	Infrastructure, resource, and facility databases	Geographical areas of damage, probabilities and numbers of fatalities, and resource allocation	Uses customized geographic information systems (GISs) to display and analyze hazard predictions	Federal, state, and local emergency managers
Source: Web page no longer available.							
Risk and Vulnerability Assessment Tool (RVAT) (National Oceanic and Atmospheric Administration [NOAA])	Florida (Brevard and Colusa counties), for demonstration purposes	Natural	Layers hazards, critical facilities, societal, economic, and environmental data on top of one another to assess risk; also includes mitigation opportunity analysis	(Not found)	Aggregation of vulnerabilities and mitigation opportunities	Vulnerability assessment: hazard identification, hazard analysis, critical facility analysis, societal analysis, economic analysis, environmental analysis, mitigation opportunity analysis, and interactive mapping	Brevard and Colusa counties, Florida; meant to be an illustrative model for other counties on how to perform risk assessment
Source: http://www.csc.noaa.gov/rvat/							

Table H.1—Continued

Name (Owner)	Access	Hazards	Scope	User Input	Output	Approach	Target Clientele
CounterMeasures Risk Analysis Software (Alion Science and Technology)	Commercial	Used to assess "security" and government regulation compliance	"Tailor-made" assessments to gauge risk	Site-specific data	Areas of risk and compliance with government regulations	Package of assessment tools used to assess security and government regulation compliance; checklist approach	Government and commercial organizations
Source: http://www.countermeasures.com							
Risk Analysis and Management for Critical Asset Protection (RAMCAP) (ASME Innovative Technologies Institute)	Commercial	Intentional: CBRNE	Assessment of threat to and consequences of intentional attacks on critical U.S. infrastructure	Specifics of information required not available	(Not found)	Part of U.S. Department of Homeland Security (DHS) national critical infrastructure assessment; includes threat assessment, cost-benefit tool, business continuity tool, and risk analysis for infrastructure interdependency	Local, state, and federal governments, as well as private-sector infrastructure stakeholders
Source: http://www.asme-iti.org/RAMCAP/RAMCAP_Plus_2.cfm							
Risk Assessment Methodology for Communities (RAM-C) (Sandia National Laboratories)	Commercial	All	Security Risk Assessment Methodology evaluating likelihood of attack, consequences, and protection-system effectiveness	Identification of threats and consequences, characterization of facilities and current system effectiveness (risk reduction capabilities)	Identification of how well people and assets are protected; determination of relative risk; identification of vulnerabilities; analysis of how those vulnerabilities can be minimized	Local community participation process involves planning, facility characterization, assessing threats and consequences, determining protection-system effectiveness, and assessing risk threshold	Local governments and other stakeholders
Source: http://www.sandia.gov/ram/RAM-C%20overview%20paper.pdf							

Table H.1—Continued

Name (Owner)	Access	Hazards	Scope	User Input	Output	Approach	Target Clientele
Probabilistic Terrorism Model (Risk Management Solutions [RMS])	Commercial	Intentional	Model used to assess the economic implications of macroterrorism (causing more than $1 million in economic losses, more than 100 fatalities or 500 injuries, or massively symbolic damage)	Relative likelihood of a particular city or target being attacked and the absolute probability of an attack or coordinated attacks	Probability of attack	Derived through expert elicitation process informed by terrorist attack history and contextual trends	Insurance companies
Source: http://www.rms.com/Terrorism/Solutions/ProbabilisticTerrorismModel.asp							
Core Vulnerabilities Assessment Management Program (CVAMP) (U.S. Navy)	U.S. Navy	All	Vulnerability assessment from known capabilities	Site assessments	Areas of vulnerability	Means to manage and prioritize vulnerabilities, as well as resources required to address the vulnerability	U.S. Navy
Source: http://www.defenselink.mil/policy/sections/policy_offices/hd/assets/downloads/dcip/education/outreach/CVAMPBriefing.pdf							
Blast/FX Explosive Effects Analysis Software (Northrop Grumman Corporation)	Commercial	Explosions	Modeling for explosions within facilities	Site-specific data	Areas and people at risk if there is an explosion	Model	Local, state, and federal government agencies; airport authorities; other Transportation Security Administration (TSA)–approved organizations
Source: http://www.blastfx.com/							

Name (Owner)	Access	Hazards	Scope	User Input	Output	Approach	Target Clientele
BASF Security Vulnerability Assessment (SVA) Methodology and Enhanced Security Implementation Management (BASF)	Corporate	Chemical	Assess the security of BASF chemical facilities	Hazard identification, consequences and target attractiveness, target classification, mitigating and aggravating factors	Post-inspection report and recommendations: security gaps, countermeasure planning	Local team, expert consultation, paper analysis, site inspections	BASF-owned or -operated fixed facilities in the United States
Source: http://www.americanchemistry.com/s_rctoolkit/sec.asp?CID=1803&DID=6738							
Geographic information system (GIS) software (ESRI)	Commercial	All	Used to map pieces of information to reveal trends or track events	Must input data beyond geographical maps	Real-time mapping of location of events (not specifically designed for emergency management but applicable especially in conjunction with other tools)	Input based	Government, emergency managers, environmental protection agencies
Source: http://www.esri.com/products/							

Table H.2
Planning Tools and Methods

Name (Owner)	Access	Hazards	Scope	User Input	Output	Approach	Target Clientele
Sync Matrix (All Hazards Management and Argonne National Laboratory)	Available to state and local governments through the Federal Emergency Management Agency (FEMA). Commercial availability is limited.[a]	All	Multi-agency planning tool, activity tracking, information warehousing	Site-specific data	Local multi-agency tool for planning, generating exercises, storing resource information	SharePoint based	Any federal, state, or local government entity conducting emergency planning
Source: http://www.scienceblog.com/community/older/2005/10/200509561.shtml							
Incident Resource Inventory System (IRIS) (U.S. Department of Homeland Security [DHS])	State, local, and tribal emergency responders	All hazards	Resources for mutual aid purposes based on mission requirements, capability of resources, and response time	Local emergency responders must enter site-specific resource data	Resource generation: availability of resources	Database	State, local, and tribal responders
Source: http://www.fema.gov/pdf/emergency/nims/nims_iris_fact_sheet.pdf							
Homeland Security Exercise and Evaluation Program (HSEEP) (U.S. Department of Homeland Security [DHS])	Public	All	Capabilities- and performance-based exercise program designed to provide national standards for local, state, and federal responders	User enters exercise-specific information based on the methodology and policy laid out	Toolkit helps coordinate, evaluate, and help plan exercises, as well as assess performance of the exercise	Guidelines, policies, and methodologies available online	Federal, state, local, and tribal responders
Source: https://hseep.dhs.gov/pages/1001_HSEEP7.aspx							

Table H.2—Continued

Name (Owner)	Access	Hazards	Scope	User Input	Output	Approach	Target Clientele
Lessons Learned Information Sharing (LLIS) (U.S. Department of Homeland Security [DHS])	Public (through registration)	All	Means of sharing lessons learned	Federal, state, local, tribal, and contracted responders can upload lessons learned from training exercises and events	Collection of resources and lessons learned on emergency management at all levels of government; includes searchable database	Web based	Federal, state, local, and tribal responders
Source: https://www.llis.dhs.gov/index.do							
Responder Knowledge Base (RKB) (U.S. Department of Homeland Security [DHS])	Public	All	Online database of emergency response equipment and products	None	Database of information on products, standards, certifications, grants, and other equipment-related information	Web based	Federal, state, local, and tribal responders
Source: https://www.rkb.us/							
Emergency Preparedness Resource Inventory (EPRI) (Agency for Healthcare Research and Quality [AHRQ]/ U.S. Department of Health and Human Services [HHS])	Public	Intentional: bioterrorist attacks	Facilitates the creation of an inventory of critical resources that would be useful in responding to a bioterrorist attack	Location, by type, and resources; hierarchy of specific resources within resource types	Inventory: automated reports for use in preparedness and planning; incident response	Web based for multiple responders to enter information about their resources; itemization of available resources and locations to be used in emergency response	Local, regional, and state government
Source: http://www.ahrq.gov/research/epri/							

Table H.2—Continued

Name (Owner)	Access	Hazards	Scope	User Input	Output	Approach	Target Clientele
Capability Based Planning Methodology and Tool (CBPMT) (Johns Hopkins University Applied Physics Laboratory)	Not currently available (commercial availability pending)	All	Outlines a methodology to map local needs to the public safety groups responsible for preparing for and responding to needs and operational capabilities	Site-specific resource data	List of all capability gaps and how each maps to capability element, emergency support function (ESF), and hazard; output can be used as a task management tool for implementation of improvement plans and designate priority, lead agency, milestones, and other items	7-step methodology for hazard characterization: (1) hazard characterization, (2) select emergency support functions per hazard, (3) select target capabilities, (4) review recommended capability elements, (5) define jurisdictional capability element goals, (6) capability gap analysis, and (7) improvement plan	Local emergency managers, as well as industries looking to become more responsive to future emergencies
Source: http://www.jhuapl.edu/ott/technologies/technology/articles/P02344.asp							
Medical Capabilities Assessment and Status Tool (M-CAST) (Navy Medicine, Office of Homeland Security)	Navy, Office of the Assistant Secretary of Defense (OASD) for Health Affairs	Intentional: chemical, biological, radiological, nuclear, or explosive (CBRNE)	Measure of readiness and preparedness of military treatment facilities	Day-to-day work assignments and assets, required CBRNE ready capabilities, contingent capabilities, reactionary capabilities	Baseline of current response capabilities and resources	Tool uses scores of program standards to determine readiness and capabilities of military treatment facilities (MTFs)	MTFs
Source: Web page no longer available.							

Table H.2—Continued

Name (Owner)	Access	Hazards	Scope	User Input	Output	Approach	Target Clientele
Preparedness for Chemical, Biological, Radiological, Nuclear, and Explosive (CBRNE) Events: Questionnaire for Health Care Facilities (Agency for Healthcare Research and Quality [AHRQ]/U.S. Department of Health and Human Services [HHS])	Public	Intentional: chemical, biological, radiological, nuclear, or explosive (CBRNE)	Questionnaire for health care facilities to fill out (for themselves) to assess readiness for CBRNE events	Capabilities and resources of each health care facility	Capability gaps: checklist of preparedness metrics	Checklist for individual hospitals to assess their preparedness; no aggregation of data	Hospitals and health care facilities
Source: http://www.ahrq.gov/prep/cbrne/							
Joint Antiterrorism (JAT) Program Manager's Guide (JAT Guide) (U.S. Army Corps of Engineers Engineer Research and Development Center)	Military only (access requires pre-certification)	Intentional (terrorism)	Provides the requirements, planning processes, templates, and tools to create an anti-terrorism (AT) program	Base-specific information	AT program	Electronic program management and decision aid to develop installation AT program	Military (all services)
Source: http://gsl.erdc.usace.army.mil/jat.html							

[a] See Argonne National Laboratory Office of Technology Transfer (undated).

Table H.3
Event Management Tools and Methods

Name (Owner)	Access	Hazards	Scope	User Input	Output	Approach	Target Clientele
WebEOC (ESi Acquisition)	Commercial (can cost >$100,000)	All	Professional version includes Status Board Suite, Chat, Checklists, Contacts, Messages, Board Wizard, Simulator, Reporter, National Weather Service (NWS) Alerts, File Library, MapTac (option), geographic information service (GIS) (option), VRiskMap Interface (option), EmerGeo Interface (option)	User-specific data entry (e.g., hospital entering number of available beds)	Real-time situational awareness, communication options, prioritization, simulator, warnings, maps	Web-based real-time crisis information management to facilitate local decisionmaking	Emergency managers

Source: http://www.esi911.com/esi/

Table H.3—Continued

Name (Owner)	Access	Hazards	Scope	User Input	Output	Approach	Target Clientele
E Team (NC4)	Commercial	All	Web-based platform providing common operating picture for coordinated incident management within and across jurisdictions	(Not found)	Standard package: incident reporting, resource and critical asset management, infrastructure status reporting, agency and jurisdiction situation reporting of disaster's impact, duty logs, planned event and activity reporting, call center tracking, vendor tracking, critical infrastructure tracking, and action planning. Enterprise package includes the above plus case management to support disaster recovery efforts, damage assessment, HAZMAT tier II reporting of detailed data on facilities that house tier II chemicals, and public information reporting for coordinating information release	Web-based incident management tools	Small cities, large municipalities, and federal agencies in various settings, including emergency operation centers, Fusion Centers, intelligence gathering and threat assessment, public health, planned event management, and training and exercises

Source: http://www.nc4.us/ETeam.php

Table H.3—Continued

Name (Owner)	Access	Hazards	Scope	User Input	Output	Approach	Target Clientele
Domestic Emergency Response Information Service (DERIS) (ICF International)	Federal Emergency Management Agency (FEMA)	All	Web-based portal to support coordinated emergency response (founded on information sharing)	(Not found)	Real-time data; video and voice capabilities (communication tools)	(Not available)	FEMA, U.S. Department of Defense (DoD)
Source: http://www.icfi.com/Markets/Defense/defense-emergency.asp#4							
Rapid Hazard Analysis (RHA) (U.S. Navy)	U.S. Navy	All	Plots potential damage of immediate threat (e.g., suspicious package)	Situation specifics	Potential consequences of immediate hazard		U.S. Navy
Source: http://damagemodels.awardspace.com/rapid_hazard.php							
Electronic Surveillance System for the Early Notification of Community-Based Epidemics (ESSENCE) (Johns Hopkins University Applied Physics Lab [JHU/APL])	Military version via Office of the Assistant Secretary of Defense for Health Affairs (OASD/HA); civilian version freely available via JHU/APL	Health events	Real-time surveillance and event detection system to support alerting of health anomalies and trigger rapid investigation	Information or data gathered at the point-of-patient-care level	Real-time syndromic surveillance (pattern detection)	Data collection; information transferred from hospital or clinic and then transferred (securely) to collection point	Military and civilian hospitals and clinics, including U.S. Department of Veterans Affairs (separate systems and access); civilian emergency managers
Source: http://www.dhss.mo.gov/ESSENCE/							
Real-Time Outbreak and Disease Surveillance (RODS) (Real-Time Outbreak and Disease Surveillance Laboratory)	Public	Health events	Real-time public health surveillance software collecting and analyzing disease surveillance data	Emergency room visit abstracts	Monitoring of increases in patients with symptoms of influenza, respiratory illness, diarrhea, and skin rashes; displays spatial distribution of cases	Aggregation of hospital and clinic data to monitor larger patterns of disease and medical care	Public health responders, hospitals, medical clinics
Source: http://openrods.sourceforge.net/							

Table H.3—Continued

Name (Owner)	Access	Hazards	Scope	User Input	Output	Approach	Target Clientele
BD RedBat (Becton, Dickinson and Company)	Commercial	Health events	Sends information from hospital or clinic to local health department; can also track statistical relevance of occurrences	Input by hospital or clinic	Real-time data on health events	Syndromic surveillance: data mining for event detection	Point of care (hospitals and clinics) and public health agencies
Source: http://www.bd.com/ds/informatics/redbat.html							
Computer Desktop Notification System (CDNS) (AtHoc)	U.S. Navy	All	Alerts in the form of pop-up windows on a desktop or mobile device accompanied by audio; includes instruction for action	None	Alert	Not applicable	U.S. Navy
Source: http://www.athoc.com/news/press-releases/2006/release-navy.html							
ePigeon Instant Messaging (ePigeon)	Commercial ~$10/user)	All hazards	Communication tool	Not applicable	Not applicable	Communication tool	Not applicable
Source: http://www.epigeon-instant-messaging.com/							

Table H.3—Continued

Name (Owner)	Access	Hazards	Scope	User Input	Output	Approach	Target Clientele
Computer-Aided Management of Emergency Operations (CAMEO), including Mapping Applications for Response, Planning, and Local Operational Tasks (MARPLOT) and Areal Locations of Hazardous Atmospheres (ALOHA) (National Oceanic and Atmospheric Administration [NOAA])	Public	Chemical	Allows planners to store information and provides easily accessible and accurate response information	Chemicals of concern	Model potential chemical releases, manage planning data: MARPLOT maps chemical events; ALOHA provides atmospheric dispersion model	Location-specific information entered by the user; information management	First responders to a chemical emergency

Source: http://response.restoration.noaa.gov/topic_subtopic_entry.php?RECORD_KEY(entry_subtopic_topic)=entry_id,subtopic_id,topic_id&entry_id(entry_subtopic_topic)=520&subtopic_id(entry_subtopic_topic)=24&topic_id(entry_subtopic_topic)=1

Table H.4
Multi-Phase Tools and Methods

Name (Owner)	Access	Hazards	Scope	User Input	Output	Approach	Target Clientele
LiveProcess (LiveProcess)	Commercial, via secure website	All	National Incident Management System (NIMS)–compliant system to support planning and response	Site-specific data	Threat identification, standardized plans (accessible anywhere), training, competency measurement, communication on and off site, compliance with NIMS, event logging, reporting, facility dashboard	Web-based platform for notification, hazard vulnerability analysis, planning, exercises, response (incident command system [ICS] mobilization and demobilization, common terminology, event logging, cross-facility cooperation and facility dashboard), information warehousing	Hospitals

Source: http://www.liveprocess.com/Our-Platform/the-liveprocess-platform

Table H.4—Continued

Name (Owner)	Access	Hazards	Scope	User Input	Output	Approach	Target Clientele
Continual Preparedness System (CPS) (Previstar)	Commercial	All	Automated National Incident Management System (NIMS)–based processes for preparedness, response, and recovery, including a searchable database for documents, personnel, equipment, maps, facilities, and supplies; tracking of resources; incident-related tasks and plans; situational awareness using geographic information systems (GISs), real-time video, and infrastructure maps; and modeling tools for hazard impacts and resource requirements (custom and pre-built models)	User-specific data	Planning generator, exercises, task lists, status boards, communication tools, inventories, needs, deployments, costs, damage assessments, modeling (biological agents and disease), debris models	Online command center with integrated package of tools spanning all phases in the disaster cycle; step-by-step proactive planning	Any emergency managers (now required across U.S. Navy)
Source: http://www.previstar.com/Products/products_CPS.asp							
Microsoft NetMeeting (Microsoft)	Public	All hazards	Communication tool	None	Communication	Live video conferencing	Anyone with the information technology (IT) to support video conferencing
Source: http://www.microsoft.com/downloads/details.aspx?FamilyID=26c9da7c-f778-4422-a6f4-efb8abba021e&displaylang=en							

Bibliography

AFI 10-245—*see* U.S. Department of the Air Force (2009).

AFI 10-802—*see* U.S. Department of the Air Force (2002).

AFI 10-2501—*see* U.S. Department of the Air Force (2007 [2009]).

AFI 41-106—*see* U.S. Department of the Air Force (2008 [2009]).

Agency for Healthcare Research and Quality, *Altered Standards of Care in Mass Casualty Events*, Rockville, Md., 2005a. As of December 16, 2009:
http://www.ahrq.gov/research/altstand/altstand.pdf

———, "Emergency Preparedness Resource Inventory (EPRI): A Tool for Local, Regional, and State Planners," Web page, current as of May 2005b. As of January 19, 2010:
http://www.ahrq.gov/research/epri/

———, *Mass Medical Care with Scarce Resources: A Community Planning Guide*, Rockville, Md., 07-0001, 2007a. As of December 16, 2009:
http://purl.access.gpo.gov/GPO/LPS80107

———, "Preparedness for Chemical, Biological, Radiological, Nuclear, and Explosive Events: Questionnaire for Health Care Facilities," current as of April 2007b. As of January 19, 2010:
http://www.ahrq.gov/prep/cbrne/

AHRQ—*see* Agency for Healthcare Research and Quality.

Alion Science and Technology, "CounterMeasures™ Risk Analysis Software," undated home page. As of January 18, 2010:
http://www.countermeasures.com/

Anderson, MAJ(P) Mike, U.S. Army, Joint Staff, Deputy Director of Antiterrorism/Homeland Defense, "Core Vulnerabilities Assessment Management Program (CVAMP) Training," undated briefing. As of January 19, 2010:
http://www.defenselink.mil/policy/sections/policy_offices/hd/assets/downloads/dcip/education/outreach/CVAMPBriefing.pdf

ANL—*see* Argonne National Laboratory.

AR 525-13—*see* U.S. Department of the Army (2008a).

AR 525-27—*see* U.S. Department of the Army (2008b).

"Argonne, All Hazards Management Announce Largest Technology License," *Science Blog*, July 2005. As of January 19, 2010:
http://www.scienceblog.com/community/older/2005/10/200509561.shtml

Argonne National Laboratory, *Introducing EMTools: The Complete Emergency Management Lifecycle Solution*, undated (a). As of December 10, 2009:
http://www.dis.anl.gov/pubs/62346.pdf

———, "Sync Matrix," Web page, undated (b). As of December 10, 2009:
http://www.dis.anl.gov/projects/syncmatrix.html

Argonne National Laboratory Office of Technology Transfer, undated home page. As of January 19, 2010:
http://www.anl.gov/techtransfer/

Arn, Mark R., *The Department of Defense's Role in Disaster Recovery*, Carlisle Barracks, Pa.: U.S. Army War College, July 4, 2006. As of December 10, 2009:
http://handle.dtic.mil/100.2/ADA461435

Arquilla, John, and David Ronfeldt, eds., *Networks and Netwars: The Future of Terror, Crime, and Militancy*, Santa Monica, Calif.: RAND Corporation, MR-1382-OSD, 2001. As of April 2009:
http://www.rand.org/pubs/monograph_reports/MR1382/

ASME Innovative Technologies Institute, "RAMCAP Plus Process," undated Web page. As of February 1, 2010:
http://www.asme-iti.org/RAMCAP/RAMCAP_Plus_2.cfm

ASPR—*see* Assistant Secretary for Preparedness and Response.

Assistant Secretary for Preparedness and Response, U.S. Department of Health and Human Services, undated home page. As of December 16, 2009:
http://www.hhs.gov/aspr/

AtHoc, "U.S. Navy Awards AtHoc with Contract to Provide Network-Centric Emergency Notification Systems Reaching up to 400,000 Personnel in Four Navy Regions," press release, Burlingame, Calif., September 12, 2006. As of January 19, 2010:
http://www.athoc.com/news/press-releases/2006/release-navy.html

Banks, David, and Kathleen Carley, "Metric Inference for Social Networks," *Journal of Classification*, Vol. 11, No. 1, March 1994, pp. 121–149.

BASF Corporation, *BASF Security Vulnerability Assessment (SVA) Methodology and Enhanced Security Implementation Management*, undated paper. As of January 19, 2010:
http://www.americanchemistry.com/s_rctoolkit/sec.asp?CID=1803&DID=6738

Beckton, Dickinson and Company, "Informatics: BD RedBat Features and Benefits," undated Web page. As of January 19, 2010:
http://www.bd.com/ds/informatics/redbat.html

Breiger, Ronald L., Kathleen M. Carley, and Philippa Pattison, *Dynamic Social Network Modeling and Analysis: Workshop Summary and Papers*, Washington, D.C.: National Academies Press, 2003.

Brown, Shona L. and Kathleen M. Eisenhardt, "The Art of Continuous Change: Linking Complexity Theory and Time-Paced Evolution in Relentlessly Shifting Organizations," *Administrative Science Quarterly*, Vol. 42, 1997, pp. 1–34.

Burt, Ronald S., *Structural Holes: The Social Structure of Competition*, Cambridge, Mass.: Harvard University Press, 1992.

Bush, George W., Organization and Operation of the Homeland Security Council, Washington, D.C.: White House Office of the Press Secretary, Homeland Security Presidential Directive 1, October 29, 2001. As of January 14, 2010:
http://www.dhs.gov/xabout/laws/gc_1213648320189.shtm

———, Management of Domestic Incidents, Washington, D.C.: White House Office of the Press Secretary, Homeland Security Presidential Directive 5, February 28, 2003a. As of December 10, 2009:
http://www.dhs.gov/xabout/laws/gc_1214592333605.shtm

———, National Preparedness, Washington, D.C.: White House Office of the Press Secretary, Homeland Security Presidential Directive 8, December 17, 2003b. As of December 10, 2009:
http://www.dhs.gov/xabout/laws/gc_1215444247124.shtm

———, Public Health and Medical Preparedness, Homeland Security Presidential Directive 21, October 18, 2007. As of December 10, 2009:
http://www.dhs.gov/xabout/laws/gc_1219263961449.shtm

Carley, Kathleen, Ju-Sung Lee, and David Krackhardt, "Destabilizing Terrorist Networks," *Connections*, Vol. 24, No. 3, 2001, pp. 79–92.

Carley, Kathleen M., Jeffrey Reminga, and Natasha Kamneva, "Destabilizing Terrorist Networks," Pittsburgh, Pa.: Center for Computational Analysis of Social and Organizational Systems, Carnegie Mellon University, 2003. As of December 10, 2009:
http://www.casos.cs.cmu.edu/publications/papers/carley_2003_destabilizingterrorist.pdf

CDC—*see* Centers for Disease Control and Prevention.

Centers for Disease Control and Prevention, "Key Facts About the Cities Readiness Initiative (CRI)," last modified April 2, 2008. As of December 16, 2009:
http://www.bt.cdc.gov/cri/facts.asp

———, "Strategic National Stockpile (SNS)," last updated March 31, 2009. As of December 16, 2009:
http://www.bt.cdc.gov/stockpile/

CEPAR—*see* Johns Hopkins Office of Critical Event Preparedness and Response.

Child, John, *Organization: A Guide to Problems and Practice*, 2nd ed., London and Hagerstown, Md.: Harper and Row, 1984.

———, *Organization: Contemporary Principles and Practice*, Malden, Mass.: Blackwell Publishing, 2005.

CNI Instruction 3440.17—*see* U.S. Department of the Navy (2006).

Columbus Consolidated Government, "Welcome from Major Jim Wetherington," undated Web page. As of December 10, 2009:
http://www.columbusga.org/mayor/welcome.htm

Conway, Michael D., Carlo D. Rizzuto, and Leigh M. Weiss, "A Better Way to Speed the Adoption of Vaccines," *McKinsey Quarterly*, August 2008.

Corbacioglu, Sitki, and Nalm Kapucu, "Organizational Learning and Self-Adaptation in Dynamic Disaster Environments," *Disasters: The Journal of Disaster Studies, Policy, and Management*, Vol. 30, No. 2, June 2006, pp. 212–233.

Davis, Paul K., *Analytic Architecture for Capabilities-Based Planning, Mission-System Analysis, and Transformation*, Santa Monica, Calif.: RAND Corporation, MR-1513-OSD, 2002. As of December 10, 2009:
http://www.rand.org/pubs/monograph_reports/MR1513/

Defense Threat Reduction Agency, "Combat Support: Programs: Joint Staff Integrated Vulnerability Assessments," undated Web page. No longer available online.

DHS—*see* U.S. Department of Homeland Security.

Diani, Mario, and Doug McAdam, eds., *Social Movements and Networks: Relational Approaches to Collective Action*, Oxford and New York: Oxford University Press, 2003.

Dixon, Lloyd, and Rachel Kaganoff Stern, *Compensation for Losses from the 9/11 Attacks*, Santa Monica, Calif.: RAND Corporation, MG-264-ICJ, 2004. As of December 10, 2009:
http://www.rand.org/pubs/monographs/MG264/

DoD—*see* U.S. Department of Defense.

DoDD 1100.20—*see* U.S. Department of Defense (2004).

DoDD 2000.12—*see* U.S. Department of Defense (2003b).

DoDD 2060.02—*see* U.S. Department of Defense (2007).

DoDD 3003.01—*see* U.S. Department of Defense (2006a).

DoDD 3025.1—*see* U.S. Department of Defense (1993).

DoDD 3025.12—*see* U.S. Department of Defense (1994).

DoDD 3025.15—*see* U.S. Department of Defense (1997).

DoDD 3025.16—*see* U.S. Department of Defense (2000).

DoDD 3160.01—*see* U.S. Department of Defense (2008).

DoDD 5525.5—*see* U.S. Department of Defense (1986).

DoDD 6200.3—*see* U.S. Department of Defense (2003a).

DoDI 2000.16—*see* U.S. Department of Defense (2006c).

DoDI 2000.18—*see* U.S. Department of Defense (2004b).

DoDI 5200.08—*see* U.S. Department of Defense (2005b).

DOJ—*see* U.S. Department of Justice.

Donaldson, Lex, *American Anti-Management Theories of Organization: A Critique of Paradigm Proliferation*, Cambridge: Cambridge University Press, 1995.

DTRA—*see* Defense Threat Reduction Agency.

Dunbar, R. I. M., "Neocortex Size as a Constraint on Group Size in Primates," *Journal of Human Evolution*, Vol. 22, No. 6, June 1992, pp. 469–493.

Eisenhardt, Kathleen M., "Building Theories from Case Study Research," *Academy of Management Review*, Vol. 14, No. 4, October 1989, pp. 532–550.

EMAC—*see* Emergency Management Assistance Compact (undated).

EMCAPS—*see* Johns Hopkins Office of Critical Event Preparedness and Response (undated).

Emergency Management Assistance Compact, undated home page. As of December 10, 2009:
http://www.emacweb.org/

Emirbayer, Mustafa, and Jeff Goodwin, "Network Analysis, Culture, and the Problem of Agency," *American Journal of Sociology*, Vol. 99, No. 6, May 1994, pp. 1411–1454.

EMP—*see* U.S. Department of Veterans Affairs (2008).

Enders, Walter, and Todd Sandler, *The Political Economy of Terrorism*, Cambridge, UK: Cambridge University Press, 2006.

Enders, Walter, and Xuejuan Su, "Rational Terrorists and Optimal Network Structure," *Journal of Conflict Resolution*, Vol. 51, No. 1, February 2007, pp. 33–57.

Engineer Research and Development Center Geotechnical and Structures Laboratory, U.S. Army Corps of Engineers, "Joint Antiterrorism Program Manager's Guide (JAT Guide)," updated December 17, 2004. As of January 19, 2010:
http://gsl.erdc.usace.army.mil/jat.html

ePigeon, undated home page. As of January 19, 2010:
http://www.epigeon-instant-messaging.com/

ESF #8—*see* U.S. Department of Health and Human Services (2008).

ESi Acquisition, *WebEOC® Professional Version 7*, undated. As of December 10, 2009:
http://www.esi911.com/esi/

ESRI, "Products," undated Web page. As of January 19, 2010:
http://www.esri.com/products/

Everett, Martin, and Stephen P. Borgatti, "Ego Network Betweenness," *Social Networks*, Vol. 27, No. 1, January 2005, pp. 31–38.

FBI—*see* Federal Bureau of Investigation.

Federal Bureau of Investigation, "Protecting America Against Terrorist Attack: A Closer Look at the FBI's Joint Terrorism Task Forces," December 1, 2004. As of December 10, 2009:
http://www.fbi.gov/page2/dec04/jttf120114.htm

Federal Emergency Management Agency, *Technical Assistance Catalog: Preparedness and Program Management Technical Assistance*, undated. As of December 10, 2009:
http://www.fema.gov/pdf/about/divisions/npd/npd_technical_assistance_catalog.pdf

———, "National Incident Management System Incident Resource Inventory System (NIMS-IRIS)," fact sheet, May 15, 2007a. As of January 19, 2010:
http://www.fema.gov/pdf/emergency/nims/nims_iris_fact_sheet.pdf

———, "Principles of Emergency Management: Supplement," September 11, 2007b. As of December 10, 2009:
http://training.fema.gov/EMIWeb/edu/08conf/Emergency%20Management%20Principles%20Monograph%20Final.doc

———, "National Incident Management System Incident Resource Inventory System (NIMS-IRIS) Update Release—Version 2.0," NIMS Alert, September 18, 2007c. As of December 2009:
http://www.fema.gov/library/file?type=publishedFile&file=iris_version2_0_nims_alert__2_.pdf&fileid=691f9660-6b91-11dc-9950-000bdba87d5b

———, *National Response Framework*, Washington, D.C. January 2008a. As of December 10, 2009:
http://www.fema.gov/pdf/emergency/nrf/nrf-core.pdf

———, *National Incident Management System*, Washington, D.C., December 18, 2008b. As of December 10, 2009:
http://www.fema.gov/pdf/emergency/nims/NIMS_core.pdf

———, "HAZUS: FEMA's Methodology for Estimating Potential Losses from Disasters," last modified October 7, 2009a. As of January 18, 2010:
http://www.fema.gov/plan/prevent/hazus/

———, "HAZMUS-MH Overview," last modified October 13, 2009b. As of December 10, 2009:
http://www.fema.gov/plan/prevent/hazus/hz_overview.shtm

———, *Responder Knowledge Base*, version 3.4.4, November 2009c. As of December 10, 2009:
https://www.rkb.us/

FEMA—*see* Federal Emergency Management Agency.

Fernandez, Roberto M., and Doug McAdam, "Social Networks and Social Movements: Multiorganizational Fields and Recruitment to Mississippi Freedom Summer," *Sociological Forum*, Vol. 3, No. 3, June 1988, pp. 357–382.

Fischhoff, Baruch, Sarah Lichtenstein, Paul Slovic, Steven L. Derby, and Ralph Keeney, *Acceptable Risk*, New York: Cambridge University Press, 1981.

Friedkin, Noah E., "Horizons of Observability and Limits of Informal Control in Organizations," *Social Forces*, Vol. 62, No. 1, September 1983, pp. 54–77.

———, *A Structural Theory of Social Influence*, New York: Cambridge University Press, 1998.

GAO—*see* U.S. Government Accountability Office.

German, Jeff, "Anti-Terrorism Fusion Center Comes Together," *Las Vegas Sun*, July 31, 2007. As of December 10, 2009:
http://www.lasvegassun.com/news/2007/jul/31/anti-terrorism-fusion-center-comes-together/

German, Michael, and Jay Stanley, *What's Wrong with Fusion Centers?* New York: American Civil Liberties Union, December 2007. As of December 10, 2009:
http://www.aclu.org/files/pdfs/privacy/fusioncenter_20071212.pdf

Gladwell, Malcolm, *The Tipping Point: How Little Things Can Make a Big Difference*, New York: Little Brown, 2000.

Granovetter, Mark, "The Strength of Weak Ties," *American Journal of Sociology*, Vol. 78, No. 6, May 1973, pp. 1360–1380.

Hagen, Guy, Dennis K. Killinger, and Richard B. Streeter, "An Analysis of Communication Networks Among Tampa Bay Economic Development Organizations," *Connections*, Vol. 20, No. 2, 1997, pp. 13–22.

Hanneman, Robert A., and Mark Riddle, *Introduction to Social Network Methods*, Riverside, Calif.: University of California, 2005. As of December 10, 2009:
http://faculty.ucr.edu/~hanneman/nettext/

Harris, Jenine K., and Bruce Clements, "Using Social Network Analysis to Understand Missouri's System of Public Health Emergency Planners," *Public Health Reports*, Vol. 122, No. 4, July–August 2007, pp. 488–498.

HD/CS—*see* U.S. Department of Defense (2005a).

Hoehn, Andrew R., Adam Grissom, David Ochmanek, David A. Shlapak, and Alan J. Vick, *A New Division of Labor: Meeting America's Security Challenges Beyond Iraq*, Santa Monica, Calif.: RAND Corporation, MG-499-AF, 2007. As of December 10, 2009:
http://www.rand.org/pubs/monographs/MG499/

Homeland Security Council, *National Strategy for Pandemic Influenza*, Washington, D.C., November 2005. As of December 10, 2009:
http://purl.access.gpo.gov/GPO/LPS64971

———, *National Strategy for Homeland Security*, October 2007. As of December 10, 2009:
http://www.dhs.gov/xlibrary/assets/nat_strat_homelandsecurity_2007.pdf

HSC—*see* Homeland Security Council.

HSPD-1—*see* Bush (2001).

HSPD-5—*see* Bush (2003a).

HSPD-8—*see* Bush (2003b).

ICF International, "Domestic Threat Assessment and Reduction," undated Web page. As of January 19, 2010:
http://www.icfi.com/Markets/Defense/defense-emergency.asp

Insurrection Act—*see* U.S. Statutes, Title 2, Section 443.

JB MDL—*see* Joint Base McGuire-Dix-Lakehurst.

JFHQ-NCR—*see* Joint Force Headquarters National Capital Region.

Johns Hopkins Office of Critical Event Preparedness and Response, "Electronic Mass Casualty Assessment and Planning Scenarios—EMCAPS," undated Web page. As of December 10, 2009:
http://www.hopkins-cepar.org/EMCAPS/EMCAPS.html

Johns Hopkins University Applied Physical Laboratory, "Capability Based Planning Methodology and Tool," last verified January 4, 2010. As of January 19, 2010:
http://www.jhuapl.edu/ott/technologies/technology/articles/P02344.asp

Joint Base McGuire-Dix-Lakehurst, "Air Force Implements Incident Management System," press release, McGuire Air Force Base, N.J., February 16, 2007. As of April 14, 2009:
http://www.mcguire.af.mil/news/story.asp?id=123041502

Joint Force Headquarters National Capital Region, undated home page. As of December 17, 2009:
http://www.jfhqncr.northcom.mil

Joint Task Force Civil Support, home page, updated January 21, 2010. As of December 17, 2009:
http://www.jtfcs.northcom.mil/

JP 1-02—*see* U.S. Joint Chiefs of Staff (2001 [2009]).

JP 3-07.2—*see* U.S. Joint Chiefs of Staff (2004).

JP 3-26—*see* U.S. Joint Chiefs of Staff (2009).

JP 3-27—*see* U.S. Joint Chiefs of Staff (2007a).

JP 3-28—*see* U.S. Joint Chiefs of Staff (2007b).

JTF-CS—*see* Joint Task Force Civil Support.

Kaiser Permanente, *Medical Center and Hazard Vulnerability Analysis*, 2001. As of December 10, 2009:
http://www.njha.com/ep/pdf/627200834041PM.pdf

Kaplan, Eben, "Fusion Centers," Council on Foreign Relations, backgrounder, February 22, 2007. As of December 10, 2009:
http://www.cfr.org/publication/12689/

Kapucu, Naim, "Interorganizational Coordination in Dynamic Contexts: Networks in Emergency Response Management," *Connections*, Vol. 26, No. 2, 2005, pp. 33–48. As of December 10, 2009: http://www.insna.org/PDF/Connections/v26/2005_I-2-5.pdf

Kasperson, Roger E., and Jeanne X. Kasperson, "The Social Amplification and Attenuation of Risk," *Annals of the American Academy of Political and Social Science*, Vol. 545, No. 1, May 1996, pp. 95–105.

Keynes, John Maynard, *A Treatise on Probability*, London: Macmillan and Co., 1921.

Knight, Frank H., *Risk, Uncertainty and Profit*, Boston and New York: Houghton Mifflin Co., 1921.

Knoke, David, and James H. Kuklinski, *Network Analysis*, Beverly Hills, Calif.: Sage Publications, 1982.

Levy, Judith A., and Bernice A. Pescosolido, eds., *Social Networks and Health*, Amsterdam and Boston: JAI, 2002.

LiveProcess, "The LiveProcess Platform," undated (a) Web page. As of January 19, 2009: http://www.liveprocess.com/Our-Platform/the-liveprocess-platform

———, "Product Features," undated (b) Web page. As of April 14, 2009: http://www.liveprocess.com/product/

Lombardo, Joseph S., "The ESSENCE II Disease Surveillance Test Bed for the National Capital Area," *Johns Hopkins APL Technical Digest*, Vol. 24, No. 4, October–December 2003, pp. 327–334. As of December 10, 2009:
http://www.jhuapl.edu/techdigest/td2404/Lombardo.pdf

Marlow, Eugene, and Patricia O'Connor Wilson, *The Breakdown of Hierarchy: Communicating in the Evolving Workplace*, Boston: Butterworth-Heinemann, 1997.

MCOE—*see* U.S. Army Maneuver Center of Excellence.

Medical Reserve Corps, home page, last updated December 16, 2009. As of December 16, 2009: http://www.medicalreservecorps.gov/HomePage

Merrari, Ariel, "Terrorism as a Strategy of Struggle: Past and Future," *Terrorism and Political Violence*, Vol. 11, No. 4, Winter 1999, pp. 52–65.

Microsoft, "Microsoft NetMeeting," undated Web page. As of January 19, 2010:
http://www.microsoft.com/downloads/
details.aspx?FamilyID=26c9da7c-f778-4422-a6f4-efb8abba021e&displaylang=en

Milgram, S., "The Small World Problem," *Psychology Today*, May 1967, pp. 60–67.

Monge, Peter R., and Noshir S. Contractor, *Theories of Communication Networks*, New York: Oxford University Press, 2003.

MRC—*see* Medical Reserve Corps.

National Oceanic and Atmospheric Administration, Coastal Services Center, "Risk and Vulnerability Assessment Tool (RVAT)," undated Web page. As of January 18, 2010:
http://www.csc.noaa.gov/rvat/

———, "CAMEO," last revised March 2, 2009. As of January 19, 2010:
http://response.restoration.noaa.gov/topic_subtopic_entry.php?RECORD_KEY(entry_subtopic_topic)=entry_id,subtopic_id,topic_id&entry_id(entry_subtopic_topic)=520&subtopic_id(entry_subtopic_topic)=24&topic_id(entry_subtopic_topic)=1

National Security Council, *National Strategy for Combating Terrorism*, Washington, D.C., September 2006. As of December 10, 2009:
http://purl.access.gpo.gov/GPO/LPS74421
or
http://georgewbush-whitehouse.archives.gov/nsc/nsct/2006/index.html

NC4, "E Team Incident Management and Reporting," undated Web page. As of January 19, 2010: http://www.nc4.us/ETeam.php

NCIS—*see* U.S. Naval Criminal Investigative Service.

NEONN—*see* Nevada Emergency Operations and Notification Network.

Nevada Emergency Operations and Notification Network, "FBI Field Intelligence Group (FIG)," undated Web page. As of December 10, 2009:
http://www.neonn.org/(X(1)S(dcmgtn45v0d5wt45sdctkc45))/default.aspx?MenuItemID=232&MenuGroup=Home+New&AspxAutoDetectCookieSupport=1

NOAA—*see* National Oceanic and Atmospheric Administration.

Northrop Grumman Corporation, "Blast/FX," undated home page. As of January 19, 2010:
http://www.blastfx.com/

NPG—*see* U.S. Department of Homeland Security (2007b).

NRF—*see* Federal Emergency Management Agency (2008a).

NSC—*see* National Security Council.

Occupational Safety and Health Administration, *OSHA Best Practices for Hospital-Based First Receivers of Victims from Mass Casualty Incidents Involving the Release of Hazardous Substances*, January 2005. As of January 18, 2010:
http://www.osha.gov/dts/osta/bestpractices/html/hospital_firstreceivers.html

Office of the Deputy Under Secretary of Defense (Installations and Environment), *Department of Defense Base Structure Report: Fiscal Year 2006 Baseline*, c. October 2005. As of January 14, 2010:
http://www.defense.gov/pubs/BSR_2006_Baseline.pdf

Office of Homeland Security, *National Strategy for Homeland Security*, Washington, D.C., July 2002. As of December 10, 2009:
http://purl.access.gpo.gov/GPO/LPS20641

OPNAV Instruction 3440.16c—*see* U.S. Department of the Navy (1995).

OPNAV Instruction 3440.17—*see* U.S. Department of the Navy (2005).

OSHA—*see* Occupational Safety and Health Administration.

Pentland, Brian T., "Organizations as Networks of Actions," in Joel A. C. Baum and Bill McKelvey, eds., *Variations in Organization Science: In Honor of Donald T. Campbell*, Thousand Oaks, Calif.: Sage Publications, 1999, pp. 237–254.

Previstar, undated (a) home page. As of April 14, 2009:
http://www.previstar.com/

———, "Continual Preparedness System (CPS)," undated (b) Web page. As of January 19, 2010:
http://www.previstar.com/Products/products_CPS.asp

Public Law 73-2, Economy Act of 1933, March 20, 1933.

Public Law 81-920, Federal Civil Defense Act of 1950, January 12, 1951.

Public Law 89-554, Freedom of Information Act, September 6, 1966.

Public Law 92-203, Alaska Native Claims Settlement Act, December 18, 1971.

Public Law 93-288, Federal Response Plan, May 22, 1974.

Public Law 93-579, Privacy Act, December 31, 1974.

Public Law 97-174, Veterans Administration and Department of Defense Health Resources Sharing Act, May 4, 1982.

Public Law 100-707, Robert T. Stafford Disaster Relief and Emergency Assistance Act, November 23, 1988.

Public Law 104-191, Health Insurance Portability and Accountability Act, August 1996.

Public Law 107-296, Homeland Security Act of 2002, November 25, 2002. As of December 16, 2009:
http://frwebgate.access.gpo.gov/cgi-bin/getdoc.cgi?dbname=107_cong_public_laws&docid=f:publ296.107.pdf

Public Law 107-297, Terrorism Risk Insurance Act of 2002, November 26, 2002. As of January 14, 2010:
http://frwebgate.access.gpo.gov/cgi-bin/getdoc.cgi?dbname=107_cong_public_laws&docid=f:publ297.107.pdf

Public Law 107-314, Bob Stump National Defense Authorization Act for Fiscal Year 2003, December 2, 2002. As of December 17, 2009:
http://frwebgate.access.gpo.gov/cgi-bin/getdoc.cgi?dbname=107_cong_public_laws&docid=f:publ314.107.pdf

Public Law 109-417, Pandemic and All-Hazards Preparedness Act, December 19, 2006. As of December 10, 2009:
http://frwebgate.access.gpo.gov/cgi-bin/getdoc.cgi?dbname=109_cong_public_laws&docid=f:publ417.109.pdf

"Rapid Hazard Models," undated, unattributed Web page. As of January 19, 2010:
http://damagemodels.awardspace.com/rapid_hazard.php

Real-Time Outbreak and Disease Surveillance Laboratory, "The RODS Open Source Project: Open Source Outbreak and Disease Surveillance Software," undated Web page. As of January 19, 2010:
http://openrods.sourceforge.net/

Renn, Ortwin, *Risk Governance: Towards an Integrative Approach*, Geneva: International Risk Governance Council, 2005.

Ressler, Steve, "Social Network Analysis as an Approach to Combat Terrorism: Past, Present, and Future Research," *Homeland Security Affairs*, Vol. II, No. 2, July 2006. As of December 10, 2009:
http://www.hsaj.org/pages/volume2/issue2/pdfs/2.2.8.pdf

Risk Management Solutions, "Probabilistic Terrorism Model," undated Web page. As of January 19, 2010:
http://www.rms.com/Terrorism/Solutions/ProbabilisticTerrorismModel.asp

Ritzer, George, ed., *Handbook of Social Problems: A Comparative International Perspective*, Thousand Oaks, Calif.: Sage Publications, 2004.

RKB—*see* Federal Emergency Management Agency (2009).

RMS—*see* Risk Management Solutions.

SAIC, Consequence Assessment Tool Set, undated Web page. No longer available.

Sandia Corporation, "Security Risk Assessment Methodology for Communities (RAM-C™)," undated paper. As of January 19, 2010:
http://www.sandia.gov/ram/RAM-C%20overview%20paper.pdf

School of the Americas Watch, "A Brief History of the SOA Watch Movement," undated. As of September 8, 2008:
http://www.soaw.org/docs/08orgpacket/19-21history.pdf

Scott, John, *Social Network Analysis: A Handbook*, 2nd ed., Thousand Oaks, Calif.: Sage Publications, 2000.

Serenko, Alexander, Nick Bontis, and Timothy Hardie, "Organizational Size and Knowledge Flow: A Proposed Theoretical Link," *Journal of Intellectual Capital*, Vol. 8, No. 4, October 2007, pp. 610–627.

Slovic, Paul, Baruch Fischhoff, and Sarah Lichtenstein, "Rating the Risks," *Environment*, Vol. 21, No. 3, April 1979, pp. 14–20, 36–39.

Smith, Kirsten P., and Nicholas A. Christakis, "Social Networks and Health," *Annual Review of Sociology*, Vol. 34, August 2008, pp. 405–429.

SOA Watch—*see* School of the Americas Watch.

Stafford Act—*see* Public Law 100-707.

State of Missouri Department of Health and Senior Services, "ESSENCE," undated Web page. As of January 19, 2010:
http://www.dhss.mo.gov/ESSENCE/

Stevens, Donald, Terry L. Schell, Thomas Hamilton, Richard Mesic, Michael Scott Brown, Edward W. Chan, Mel Eisman, Eric V. Larson, Marvin Schaffer, Bruce Newsome, John Gibson, and Elwyn Harris, *Near-Term Options for Improving Security at Los Angeles International Airport*, Santa Monica, Calif.: RAND Corporation, DB-468-1-LAWA, 2004. As of February 1, 2010:
http://www.rand.org/pubs/documented_briefings/DB468-1/

Tierney, Kathleen J., "Emergency Medical Preparedness and Response in Disasters: The Need for Interorganizational Coordination," *Public Administration Review*, Vol. 45, January 1985, pp. 77–84.

U.S. Army Maneuver Center of Excellence, "Fact Sheet 2009," 2009. As of December 10, 2009: https://www.benning.army.mil/mcoe/content/publications/fact%20sheet2009.pdf

U.S. Code, Title 10, Armed Forces, Chapter 15, Enforcement of the Laws to Restore Public Order, Section 331, Federal Aid for State Governments. As of December 10, 2009: http://frwebgate.access.gpo.gov/cgi-bin/usc.cgi?ACTION=RETRIEVE&FILE=$$xa$$busc10.wais&start=1331432&SIZE=1863&TYPE=TEXT

———, Title 10, Armed Forces, Chapter 15, Enforcement of the Laws to Restore Public Order, Section 332, Use of Militia and Armed Forces to Enforce Federal Authority. As of December 10, 2009: http://frwebgate.access.gpo.gov/cgi-bin/usc.cgi?ACTION=RETRIEVE&FILE=$$xa$$busc10.wais&start=1333301&SIZE=6613&TYPE=TEXT

———, Title 10, Armed Forces, Chapter 15, Enforcement of the Laws to Restore Public Order, Section 333, Major Public Emergencies; Interference with State and Federal Law. As of December 10, 2009: http://frwebgate.access.gpo.gov/cgi-bin/usc.cgi?ACTION=RETRIEVE&FILE=$$xa$$busc10.wais&start=1339920&SIZE=4203&TYPE=TEXT

———, Title 10, Armed Forces, Chapter 15, Enforcement of the Laws to Restore Public Order, Section 334, Proclamation to Disperse. As of December 10, 2009: http://frwebgate.access.gpo.gov/cgi-bin/usc.cgi?ACTION=RETRIEVE&FILE=$$xa$$busc10.wais&start=1344129&SIZE=5578&TYPE=TEXT

———, Title 10, Armed Forces, Chapter 15, Enforcement of the Laws to Restore Public Order, Section 335, Guam and Virgin Islands Included as "State." As of December 10, 2009: http://frwebgate.access.gpo.gov/cgi-bin/usc.cgi?ACTION=RETRIEVE&FILE=$$xa$$busc10.wais&start=1349713&SIZE=1495&TYPE=TEXT

———, Title 10, Armed Forces, Chapter 18, Military Support for Civilian Law Enforcement Agencies, Section 371, Use of Information Collected During Military Operations. As of December 10, 2009: http://frwebgate.access.gpo.gov/cgi-bin/usc.cgi?ACTION=RETRIEVE&FILE=$$xa$$busc10.wais&start=1358585&SIZE=3671&TYPE=TEXT

———, Title 10, Armed Forces, Chapter 18, Military Support for Civilian Law Enforcement Agencies, Section 372, Use of Military Equipment and Facilities. As of December 10, 2009: http://frwebgate.access.gpo.gov/cgi-bin/usc.cgi?ACTION=RETRIEVE&FILE=$$xa$$busc10.wais&start=1362262&SIZE=3478&TYPE=TEXT

———, Title 10, Armed Forces, Chapter 18, Military Support for Civilian Law Enforcement Agencies, Section 373, Training and Advising Civilian Law Enforcement Officials. As of December 10, 2009: http://frwebgate.access.gpo.gov/cgi-bin/usc.cgi?ACTION=RETRIEVE&FILE=$$xa$$busc10.wais&start=1365746&SIZE=2071&TYPE=TEXT

———, Title 10, Armed Forces, Chapter 18, Military Support for Civilian Law Enforcement Agencies, Section 374, Maintenance and Operation of Equipment. As of December 10, 2009: http://frwebgate.access.gpo.gov/cgi-bin/usc.cgi?ACTION=RETRIEVE&FILE=$$xa$$busc10.wais&start=1367823&SIZE=26557&TYPE=TEXT

———, Title 10, Armed Forces, Chapter 18, Military Support for Civilian Law Enforcement Agencies, Section 375, Restriction on Direct Participation by Military Personnel. As of December 10, 2009: http://frwebgate.access.gpo.gov/cgi-bin/usc.cgi?ACTION=RETRIEVE&FILE=$$xa$$busc10.wais&start=1394386&SIZE=2212&TYPE=TEXT

———, Title 10, Armed Forces, Chapter 18, Military Support for Civilian Law Enforcement Agencies, Section 376, Support Not to Affect Adversely Military Preparedness. As of December 10, 2009: http://frwebgate.access.gpo.gov/cgi-bin/usc.cgi?ACTION=RETRIEVE&FILE=$$xa$$busc10.wais&start=1396604&SIZE=1871&TYPE=TEXT

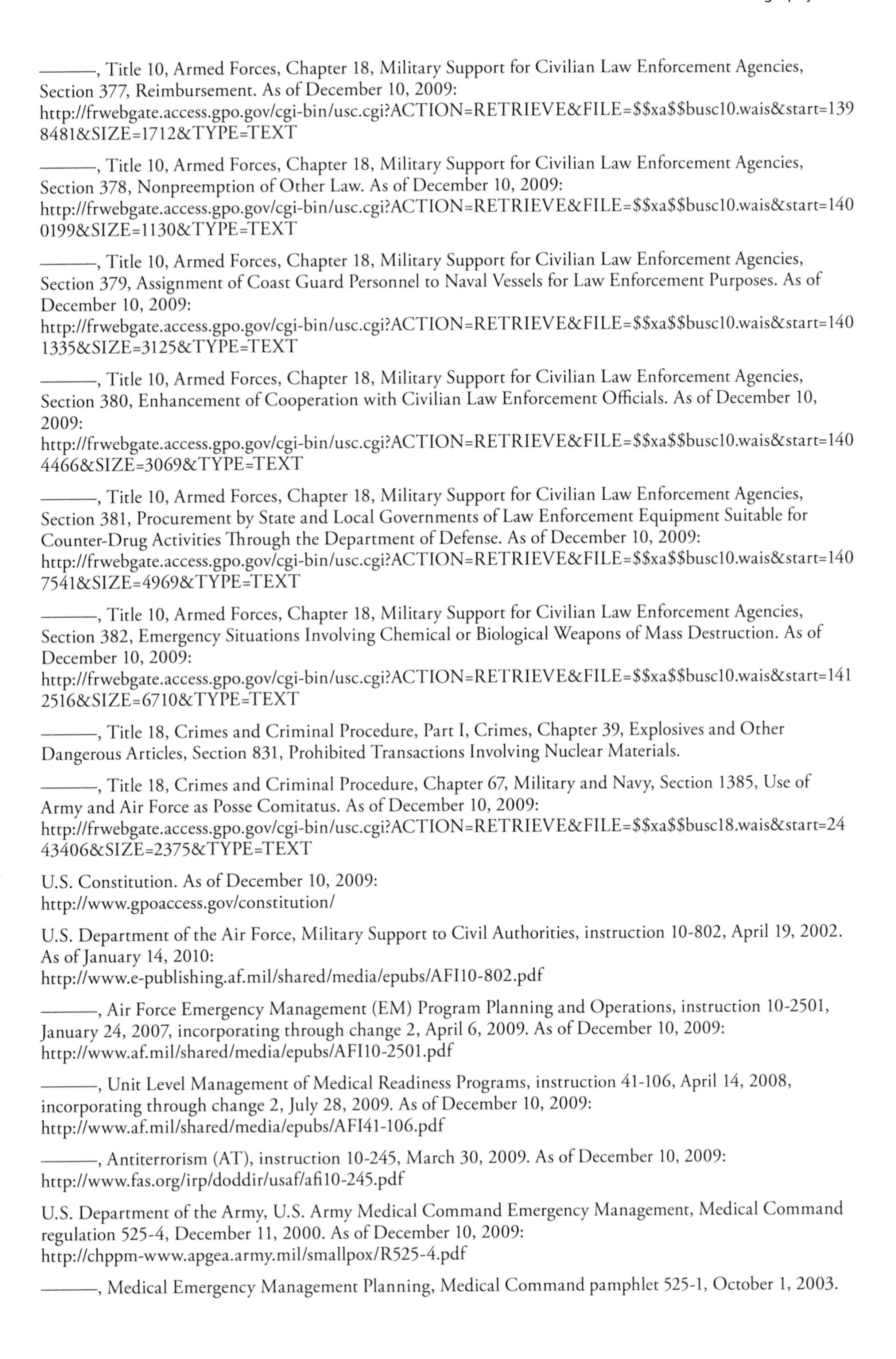

———, Title 10, Armed Forces, Chapter 18, Military Support for Civilian Law Enforcement Agencies, Section 377, Reimbursement. As of December 10, 2009:
http://frwebgate.access.gpo.gov/cgi-bin/usc.cgi?ACTION=RETRIEVE&FILE=$$xa$$busc10.wais&start=1398481&SIZE=1712&TYPE=TEXT

———, Title 10, Armed Forces, Chapter 18, Military Support for Civilian Law Enforcement Agencies, Section 378, Nonpreemption of Other Law. As of December 10, 2009:
http://frwebgate.access.gpo.gov/cgi-bin/usc.cgi?ACTION=RETRIEVE&FILE=$$xa$$busc10.wais&start=1400199&SIZE=1130&TYPE=TEXT

———, Title 10, Armed Forces, Chapter 18, Military Support for Civilian Law Enforcement Agencies, Section 379, Assignment of Coast Guard Personnel to Naval Vessels for Law Enforcement Purposes. As of December 10, 2009:
http://frwebgate.access.gpo.gov/cgi-bin/usc.cgi?ACTION=RETRIEVE&FILE=$$xa$$busc10.wais&start=1401335&SIZE=3125&TYPE=TEXT

———, Title 10, Armed Forces, Chapter 18, Military Support for Civilian Law Enforcement Agencies, Section 380, Enhancement of Cooperation with Civilian Law Enforcement Officials. As of December 10, 2009:
http://frwebgate.access.gpo.gov/cgi-bin/usc.cgi?ACTION=RETRIEVE&FILE=$$xa$$busc10.wais&start=1404466&SIZE=3069&TYPE=TEXT

———, Title 10, Armed Forces, Chapter 18, Military Support for Civilian Law Enforcement Agencies, Section 381, Procurement by State and Local Governments of Law Enforcement Equipment Suitable for Counter-Drug Activities Through the Department of Defense. As of December 10, 2009:
http://frwebgate.access.gpo.gov/cgi-bin/usc.cgi?ACTION=RETRIEVE&FILE=$$xa$$busc10.wais&start=1407541&SIZE=4969&TYPE=TEXT

———, Title 10, Armed Forces, Chapter 18, Military Support for Civilian Law Enforcement Agencies, Section 382, Emergency Situations Involving Chemical or Biological Weapons of Mass Destruction. As of December 10, 2009:
http://frwebgate.access.gpo.gov/cgi-bin/usc.cgi?ACTION=RETRIEVE&FILE=$$xa$$busc10.wais&start=1412516&SIZE=6710&TYPE=TEXT

———, Title 18, Crimes and Criminal Procedure, Part I, Crimes, Chapter 39, Explosives and Other Dangerous Articles, Section 831, Prohibited Transactions Involving Nuclear Materials.

———, Title 18, Crimes and Criminal Procedure, Chapter 67, Military and Navy, Section 1385, Use of Army and Air Force as Posse Comitatus. As of December 10, 2009:
http://frwebgate.access.gpo.gov/cgi-bin/usc.cgi?ACTION=RETRIEVE&FILE=$$xa$$busc18.wais&start=2443406&SIZE=2375&TYPE=TEXT

U.S. Constitution. As of December 10, 2009:
http://www.gpoaccess.gov/constitution/

U.S. Department of the Air Force, Military Support to Civil Authorities, instruction 10-802, April 19, 2002. As of January 14, 2010:
http://www.e-publishing.af.mil/shared/media/epubs/AFI10-802.pdf

———, Air Force Emergency Management (EM) Program Planning and Operations, instruction 10-2501, January 24, 2007, incorporating through change 2, April 6, 2009. As of December 10, 2009:
http://www.af.mil/shared/media/epubs/AFI10-2501.pdf

———, Unit Level Management of Medical Readiness Programs, instruction 41-106, April 14, 2008, incorporating through change 2, July 28, 2009. As of December 10, 2009:
http://www.af.mil/shared/media/epubs/AFI41-106.pdf

———, Antiterrorism (AT), instruction 10-245, March 30, 2009. As of December 10, 2009:
http://www.fas.org/irp/doddir/usaf/afi10-245.pdf

U.S. Department of the Army, U.S. Army Medical Command Emergency Management, Medical Command regulation 525-4, December 11, 2000. As of December 10, 2009:
http://chppm-www.apgea.army.mil/smallpox/R525-4.pdf

———, Medical Emergency Management Planning, Medical Command pamphlet 525-1, October 1, 2003.

———, Antiterrorism, regulation 525-13, September 11, 2008a.

———, Army Emergency Management Program, regulation 525-27, December 4, 2008b. As of December 11, 2009:
http://www.fas.org/irp/doddir/army/ar525-27.pdf

U.S. Department of Defense, DoD Cooperation with Civilian Law Enforcement Officials, Washington, D.C., directive 5525.5, January 15, 1986. As of December 10, 2009:
http://www.dtic.mil/whs/directives/corres/html/552505.htm

———, Military Support to Civil Authorities (MSCA), Washington, D.C., directive 3025.1, January 15, 1993. As of December 10, 2009:
http://www.dtic.mil/whs/directives/corres/html/302501.htm

———, Military Assistance for Civil Disturbances (MACDIS), Washington, D.C., directive 3025.12, February 4, 1994. As of December 10, 2009:
http://www.dtic.mil/whs/directives/corres/html/302512.htm

———, Military Assistance to Civil Authorities, Washington, D.C., directive 3025.15, February 18, 1997. As of December 10, 2009:
http://www.dtic.mil/whs/directives/corres/html/302515.htm

———, Military Emergency Preparedness Liaison Officer (EPLO) Program, Washington, D.C., directive 3025.16, December 18, 2000. As of December 10, 2009:
http://www.dtic.mil/whs/directives/corres/html/302516.htm

———, Department of Defense Installation Chemical, Biological, Radiological, Nuclear and High-Yield Explosive Emergency Response Guidelines, Washington, D.C., instruction 2000.18, December 4, 2002. As of December 10, 2009:
http://www.dtic.mil/whs/directives/corres/html/200018.htm

———, Emergency Health Powers on Military Installations, Washington, D.C., directive 6200.3, May 12, 2003a. As of December 10, 2009:
http://www.dtic.mil/whs/directives/corres/html/620003.htm

———, DoD Antiterrorism (AT) Program, Washington, D.C., directive 2000.12, August 18, 2003b. As of December 10, 2009:
http://www.dtic.mil/whs/directives/corres/pdf/200012p.pdf

———, Support and Services for Eligible Organizations and Activities Outside the Department of Defense, Washington, D.C., directive 1100.20, April 12, 2004. As of December 10, 2009:
http://www.dtic.mil/whs/directives/corres/html/110020.htm

———, *Strategy for Homeland Defense and Civil Support*, Washington, D.C., June 2005a. As of December 10, 2009:
http://purl.access.gpo.gov/GPO/LPS61327

———, Security of DoD Installations and Resources, Washington, D.C., instruction 5200.08, December 10, 2005b. As of December 10, 2009:
http://www.dtic.mil/whs/directives/corres/html/520008.htm

———, DoD Support to Civil Search and Rescue (SAR), Washington, D.C., directive 3003.01, January 20, 2006a. As of December 10, 2009:
http://www.dtic.mil/whs/directives/corres/html/300301.htm

———, *Report of the Quadrennial Defense Review*, Washington, D.C.: Secretary of Defense, February 6, 2006b. As of December 11, 2009:
http://www.defense.gov/qdr/report/Report20060203.pdf

———, DoD Antiterrorism (AT) Standards, Washington, D.C., instruction 2000.16, October 2, 2006c, incorporating through change 2, December 8, 2006. As of December 10, 2009:
http://www.dtic.mil/whs/directives/corres/html/200016.htm

———, Department of Defense (DoD) Combating Weapons of Mass Destruction (WMD) Policy, Washington, D.C., directive 2060.02, April 19, 2007. As of December 10, 2009:
http://www.dtic.mil/whs/directives/corres/pdf/206002p.pdf

———, Homeland Defense Activities Conducted by the National Guard, Washington, D.C., directive 3160.01, August 25, 2008. As of December 10, 2009:
http://www.dtic.mil/whs/directives/corres/pdf/316001p.pdf

U.S. Department of Defense, Defense Critical Infrastructure Program, "Risk Assessment," undated Web page. As of January 18, 2010:
http://policy.defense.gov/sections/policy_offices/hd/offices/dcip/risk/risk_assessment/

U.S. Department of Health and Human Services, *Emergency Support Function #8: Public Health and Medical Services Annex*, January 2008. As of December 16, 2009:
http://www.fema.gov/pdf/emergency/nrf/nrf-esf-08.pdf

U.S. Department of Homeland Security, "HSEEP Mission," Web page, undated (a). As of December 10, 2009:
https://hseep.dhs.gov/pages/1001_HSEEP7.aspx

———, "Lessons Learned Information Sharing," Web page, undated (b). As of December 10, 2009:
https://www.llis.dhs.gov/index.do

———, Office for Domestic Preparedness, *Vulnerability Assessment Methodologies Report: Phase I Final Report*, Washington, D.C., 2003a.

———, *The National Strategy for the Physical Protection of Critical Infrastructures and Key Assets*, Washington, D.C., February 2003b. As of December 10, 2009:
http://purl.access.gpo.gov/GPO/LPS28728

———, *The National Strategy to Secure Cyberspace*, Washington, D.C., February 2003c. As of December 10, 2009:
http://purl.access.gpo.gov/GPO/LPS28730

———, *Survey of Vulnerability Assessment Models*, draft, Washington, D.C., October 2003d.

———, *The National Strategy for Maritime Security*, September 2005. As of December 10, 2009:
http://www.dhs.gov/xlibrary/assets/HSPD13_MaritimeSecurityStrategy.pdf

———, *Universal Task List*, Washington, D.C., February 2007a.

———, *National Preparedness Guidelines*, Washington, D.C., September 2007b. As of December 10, 2009:
http://www.dhs.gov/xlibrary/assets/National_Preparedness_Guidelines.pdf

———, *Target Capabilities List: A Companion to the National Preparedness Guidelines*, Washington, D.C., September 2007c. As of December 10, 2009:
https://www.llis.dhs.gov/displayContent?contentID=26724

———, *National Infrastructure Protection Plan: Partnering to Enhance Protection and Resiliency*, Washington, D.C., 2009. As of December 10, 2009:
http://www.dhs.gov/xlibrary/assets/NIPP_Plan.pdf

U.S. Department of Justice, "September 11th Victim Compensation Fund of 2001," undated Web page. As of January 14, 2010:
http://www.justice.gov/archive/victimcompensation/

———, Office of Justice Programs, Department of Homeland Security, and White House Office of the Press Secretary, *Interim National Preparedness Goal: Homeland Security Presidential Directive 8—National Preparedness*, Washington, D.C., 2005.

U.S. Department of the Navy, Naval Civil Emergency Management Program, Office of the Chief of Naval Operations instruction 3440.16c, March 10, 1995.

———, Navy Installation Emergency Management Program, Office of the Chief of Naval Operations instruction 3440.17, July 22, 2005. As of December 10, 2009:
http://doni.daps.dla.mil/Directives/03000%20Naval%20Operations%20and%20Readiness/03-400%20Nuclear,%20Biological%20and%20Chemical%20Program%20Support/3440.17.pdf

———, Navy Installation Emergency Management Program Manual, Commander, Navy Installations, instruction 3440.17, January 23, 2006. As of December 10, 2009:
https://portal.navfac.navy.mil/portal/page/portal/navfac/navfac_ww_pp/navfac_nfesc_pp/amphibious%20and%20expeditionary/em-cbrn%20defense%20support/emergency%20management%20program

U.S. Department of Veterans Affairs, *Emergency Management Program Guidebook*, Washington, D.C., February 2002.

———, *Emergency Management Program Guidebook* (draft), Washington, D.C., March 2008.

U.S. Government Accountability Office, *Homeland Security: Federal Efforts Are Helping to Alleviate Some Challenges Encountered by State and Local Information Fusion Centers*, Washington, D.C., GAO-08-35, October 2007. As of December 10, 2009:
http://purl.access.gpo.gov/GPO/LPS90145

U.S. Joint Chiefs of Staff, Joint Doctrine Division, *Department of Defense Dictionary of Military and Associated Terms*, Joint Publication 1-02, Washington, D.C., April 12, 2001, as amended through August 19, 2009. As of December 10, 2009:
http://www.dtic.mil/doctrine/dod_dictionary/

———, *Joint Tactics, Techniques and Procedures for Antiterrorism*, Washington, D.C., second draft, December 8, 2004. Not available to the general public.

———, *Homeland Defense*, Washington, D.C., joint publication 3-27, July 12, 2007a. As of December 10, 2009:
http://www.dtic.mil/doctrine/new_pubs/jp3_27.pdf

———, *Civil Support*, Washington, D.C., joint publication 3-28, September 14, 2007b. As of December 10, 2009:
http://www.dtic.mil/doctrine/new_pubs/jp3_28.pdf

———, *Counterterrorism*, Washington, D.C., joint publication 3-26, November 13, 2009. As of December 17, 2009:
http://www.dtic.mil/doctrine/new_pubs/jp3_26.pdf

U.S. Marine Corps, "Chemical Biological Incident Response Force," undated Web page. As of February 1, 2010:
http://www.marines.mil/unit/cbirf/Pages/default.aspx

U.S. Naval Criminal Investigative Service, "About NCIS," undated Web page. As of June 13, 2009:
http://www.ncis.navy.mil/about.asp

U.S. Northern Command, "About USNORTHCOM," undated Web page (a). As of February 1, 2010:
http://www.northcom.mil/About/index.html

———, "CCMRF," undated Web page (b). As of December 17, 2009:
http://www.northcom.mil/News/2008/CCMRF/index.html

U.S. Pacific Command, "Joint Interagency Task Force West (JIATF-W)," last updated September 23, 2008. As of January 27, 2010:
http://www.pacom.mil/staff/jiatfwest/

U.S. Statutes, Title 2, Section 443, Insurrection Act of 1807.

USMC—*see* U.S. Marine Corps.

USNORTHCOM—*see* U.S. Northern Command.

USPACOM—*see* U.S. Pacific Command.

UTL—*see* U.S. Department of Homeland Security (2007a).

VA—*see* U.S. Department of Veterans Affairs.

Varda, Danielle M., Anita Chandra, Stefanie A. Stern, and Nicole Lurie, "Core Dimensions of Connectivity in Public Health Collaboratives," *Journal of Public Health Management and Practice*, Vol. 14, No. 5, September–October 2008, pp. E1–E7.

Wasserman, Stanley, and Katherine Faust, *Social Network Analysis: Methods and Applications*, New York: Cambridge University Press, 1994.

Wellman, Barry, and Stephen D. Berkowitz, eds., *Social Structures: A Network Approach*, New York: Cambridge University Press, 1988.

White House Office, *National Strategy to Combat Weapons of Mass Destruction*, Washington, D.C., December 2002. As of December 16, 2009:
http://purl.access.gpo.gov/GPO/LPS24899

———, *National Strategy for Information Sharing: Successes and Challenges in Improving Terrorism-Related Information Sharing*, Washington, D.C., October 2007.

Willis, Henry H. Tom LaTourrette, Terrence K. Kelly, Scot Hickey, and Samuel Neill, *Terrorism Risk Modeling for Intelligence Analysis and Infrastructure Protection*, Santa Monica, Calif.: RAND Corporation, TR-386-DHS, 2007. As of December 10, 2009:
http://www.rand.org/pubs/technical_reports/TR386/

Willis, Henry H., Andrew R. Morral, Terrence K. Kelly, and Jamison Jo Medby, *Estimating Terrorism Risk*, Santa Monica, Calif.: RAND Corporation, MG-388-RC, 2005. As of December 10, 2009:
http://www.rand.org/pubs/monographs/MG388/

Willis, Henry H., Christopher Nelson, Shoshana R. Shelton, Andrew M. Parker, John A. Zambrano, Edward W. Chan, Jeffrey Wasserman, and Brian A. Jackson, *Initial Evaluation of the Cities Readiness Initiative*, Santa Monica, Calif.: RAND Corporation, TR-640-CDC, 2009. As of December 10, 2009:
http://www.rand.org/pubs/technical_reports/TR640/

Willis, Henry H., and David S. Ortiz, *Evaluating the Security of the Global Containerized Supply Chain*, Santa Monica, Calif.: RAND Corporation, TR-214-RC, 2004. As of December 10, 2009:
http://www.rand.org/pubs/technical_reports/TR214/